THE STANDARD DE

THE ZANY WORLD OF
BASIC MATH
STUDY SIDEKICK
(1ST EDITION)

Written by The Standard Deviants® Academic Team, including:
Dr. Richard Semmler, Ph.D.
C. Alan Canant
Karen Hansen

Edited by:
Rachel Galvin
Dr. Richard Semmler, Ph.D.

Graphic Design:
C. Christopher Stevens
Sarah Fry

800-238-9669
e-mail: cerebellum@mindspring.com
www.cerebellum.com

STANDARD
DEVIANTS ™

OTHER SUBJECTS FROM CEREBELLUM:

Printed in the beautiful U.S.A.

HOW TO USE THIS BOOK

 **CHECK OUT THE VIDEO!** This Standard Deviants Study Sidekick corresponds to our Standard Deviants Video Course Review, *The Zany World of Basic Math*. Your studying and reviewing will be much more effective if you watch the *The Zany World of Basic Math* in conjunction with this workbook.

 **FOLLOW ALONG.** The **VIDEO NOTES** section does your work for you! We've already taken all of your notes—all you have to do is follow along with the video. We've even given you a **VIDEO TIME CODE**. Just reset your VCR counter to 0:00:00 when the Cerebellum logo appears at the beginning of the tape. These clocks 0:00:00 give you the time code for each important section so you know where to fast-forward to! This will enable you to learn and retain material much more efficiently. Just pause the tape after a difficult section and read through your notes!

 **LEARN NEW STUFF.** Unfortunately, we just can't include everything about Basic Math in one video. The **OTHER IMPORTANT STUFF** section gives you other cool facts you'll need to ace your tests.

TEST YOURSELF. Quizzes and Practice Exams allow you to test yourself and make sure you've covered all the bases. The answers appear at the back–*don't cheat!*

HAVE FUN. The workbook is chock-full of diversions and stress-relievers, and there's two of those neat flippy pictures on the bottom of each page.

TABLE OF CONTENTS

VIDEO TIME CODE

VIDEO NOTES

STUDY SIDEKICK

TEST YOURSELF

OTHER IMPORTANT STUFF

TABLE OF CONTENTS

VIDEO TIME CODE

VIDEO NOTES

VIDEO NOTES

Integers

`0:2:50`

We use lots of different kinds of **numbers** in math. The most common type of number is an integer. **Integers** include whole numbers and the negatives of whole numbers. This means that integers can be positive, negative, or zero. So what's the difference between whole numbers and integers? Well, not much. But **whole numbers** can only be positive, while integers can be negative. Neither whole numbers nor integers can be fractions or decimals. So 5, 134, −16, and 2,000,000 are all integers, whereas $\frac{1}{2}$ and .67 are not.

The easiest way to visualize integers is to put them on a number line. A **number line** is a vertical or horizontal line that is

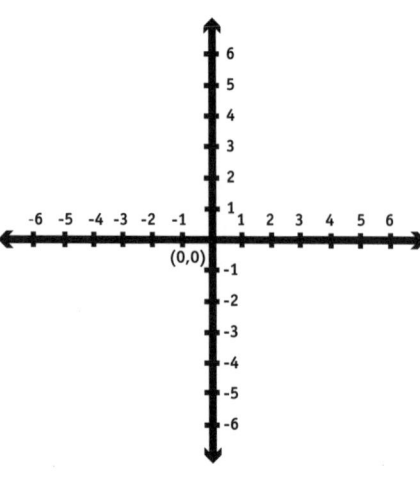

marked at even intervals, or units, similar to a thermometer. These units are labeled with numbers.

Here's how a number line works. Moving to the right or up on the number line, the numbers become greater, while moving to the left or down, the numbers become smaller. So, any number on the

number line above or to the right of zero is positive, or greater than zero. Any number below, or to the left of zero, is negative, or less than zero. This number line will be a useful tool when we get to operations on negative numbers.

The ends of a number line go on forever because there is no end to the amount of numbers we can produce. Another way of saying this is that numbers go on to infinity, kinda like your student loans.

All integers are made up of **digits**. Digits are simply the numbers 0 to 9. Every digit in a number goes in a certain digit place.

$$45,678,342$$

Look at the number above. The first digit in any whole number, (the 2 in this case) is in the **units place**. The **tens place** is just next door, on the left. One place over from the tens place is the **hundreds place**. To the left of the hundreds place, we've got the **thousands place**, the ten-thousands place, the hundred-thousands place, the millions place, the ten-millions place, and so on.

IGOR

and here's directions to Igor's place

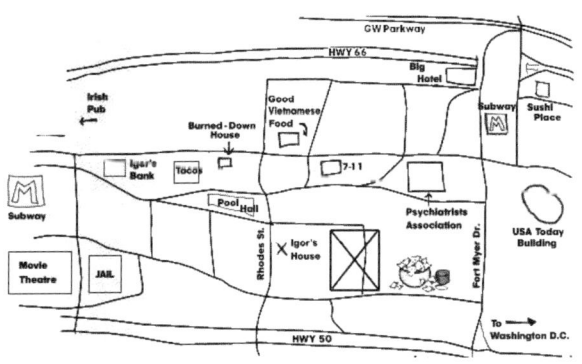

11

Just like we saw for the integers on a number line, these digit places go on forever because numbers go to infinity.

Let's take a look at the number one thousand, four hundred, ninety-two.

1492

For this number, the 2 is in the units place. Moving to the left, the 9 is in the tens place, and the 4 is in the hundreds place. Finally, the 1 is in the thousands place.

NEGATIVE

ZERO

POSITIVE
A.K.A.
WHOLE NUMBERS

Section A: Addition

`0:7:15`

Now that we know how integers are set up, we can begin to operate with them. The first operation we'll look at is **addition**.

Adding is putting things together, then counting how much you have. Terms such as "and" or "**plus**" are often used to show that an operation is addition. When you add positive integers, or whole numbers, you combine them to form a larger number. This larger number you get when you add numbers together is called a **sum**.

`0:7:20`

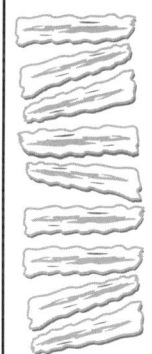

For example, if you have 4 pieces of tripe and you're lucky enough that someone gives you 5 more pieces of tripe, then you'll have 9 pieces of tripe in all. The sum of 4 pieces of tripe and 5 pieces of tripe is 9 pieces of tripe.

4 pieces of tripe + 5 pieces of tripe = 9 pieces of tripe

A Suggestion From the Standard Deviants:

Next time you're in an elevator, try walking through the wall with no door. Then, when you get out of your coma, laugh with your friends about your light-hearted tomfoolery.

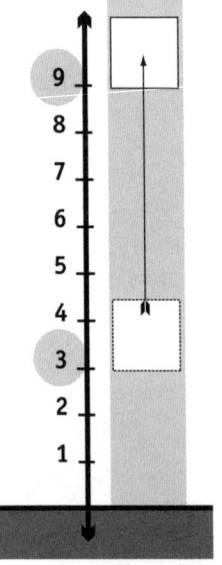

Both the operation of addition and the operation of subtraction can be best shown by using a number line.

When adding a positive integer to any other integer, you will be moving up the number line because moving up the line is an increase.

A number line is a lot like an elevator. For 3 + 6 you will start at the 3rd floor on the elevator and move up 6 floors. So after moving up 6 floors you're at the 9th floor.

3 floors plus 6 more puts you on the 9th floor.

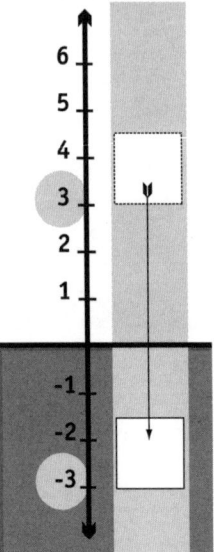

It's just the opposite when adding a negative number, like −4, −5,000, and so on.

When adding a negative integer to any other integer, you will be moving down the number line. That's because the further you move down the number line, the numbers will decrease.

It's just like when you're in an elevator and the number of the floor you're on decreases as you go down. So adding a negative number is the same as subtraction.

cerebellum
CORPORATION

Let's go back to the example where we added 3 and 6. This time we'll add 3 and −6. You still start at 3 on the number line, or the 3rd floor, then move down 6 floors.

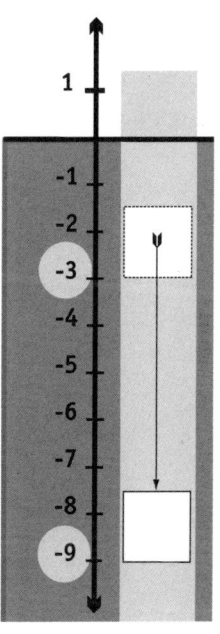

After moving down 6 floors, you're at −3. So we know that

$$3 + (-6) = -3$$

Likewise, to add −3 and −6, start at −3 and move down 6 units.

You end up at −9.

$$-3 + (-6) = -9$$

Now let's add some larger numbers.

We'll add the number of female and male ants that are in this colony. There are five hundred, sixty-seven male ants and seven hundred, eighty-nine female ants. To add big numbers like this, we need to stack the numbers on top of one another.

$$\begin{array}{r} 567 \\ + \underline{789} \end{array}$$

Important Addition Info: You always begin adding numbers on the farthest right.

Instead of chewing gum, chew bacon!

– Dr. Nick Rivers, *The Simpsons*

0:10:30

15

After we've stacked the numbers on top of one another, we combine each of the numbers in each digit place. Start by adding the numbers in the units place. In this problem, the 7 and 9 are in the units place. No problem. We just add 'em up.

$$7 + 9 = 16$$

0:12:02

Okay, we've hit a snag. Every digit place has room for only one digit, but 16 is made up of two digits.

$$
\begin{array}{r}
567 \\
+\ 789 \\
\hline
16
\end{array}
$$

Remember, digits are just the numbers 0 to 9. When you add two numbers together to fill one digit place, and the numbers add up to more than 9, then you have to carry over. **Carrying over** means you take the digit in the tens place of the two numbers you're adding together, and carry it over to the next digit place to the left. You do this whenever the sum in a digit place is greater than 9. So the number in the tens place of any digit place sum must be carried over. After we carry that digit to the next place, we include that digit in our sum for the next digit place. Let's go back to our example.

$$
\begin{array}{r}
567 \\
+\ 789 \\
\hline
16
\end{array}
$$

The 7 and 9 in the units place equal 16. So we have to carry over the 1 in the tens place of 16 to the tens place of our addition problem. We leave the 6 in the units place.

$$
\begin{array}{r}
\overset{1}{5}67 \\
+\ 789 \\
\hline
6
\end{array}
$$

Now that we've added the numbers in one digit place, we go on to the next digit place on the left, the tens place. We add the 6 and 8 in the tens place plus the 1 we carried over.

$$1 + 6 + 8 = 15$$

The tens place also has room for only one number but 15 has two. So the 1 from the tens place in 15 must be carried over to the hundreds place in our problem. This means the 5 is left in the tens place, all by itself.

$$
\begin{array}{r}
\overset{11}{5}67 \\
+\ 789 \\
\hline
56
\end{array}
$$

We'll continue from right to left for each digit place in the problem. Now we have 1, 5, and 7 in the hundreds place, so we add up all of these numbers to get the hundreds-place digit.

$$1 + 5 + 7 = 13$$

Important Dinner Party Info: Whenever you're doing a place setting, spoons always go on the farthest right.

The 3 in 13 stays in the hundreds place and the 1 goes in the thousands place. In this instance, there are no other numbers in the thousands place, so the 1 is by itself.

$$\begin{array}{r} 11 \\ 567 \\ + \ 789 \\ \hline 1{,}356 \end{array}$$

The result of adding five hundred, sixty-seven ants and seven hundred, eighty-nine ants is one thousand, three hundred, fifty-six ants. No problemo, right?

Quick Reminder:
Remember that when adding whole numbers or integers, you always begin with the numbers on the farthest right of the addition problem.

Also remember that when you add together numbers in a digit place whose sum is larger than 9, you'll have to carry over.

Section B: Subtraction

The next operation with integers we'll tackle is subtraction. **Subtraction** is the inverse, or reverse, operation of addition. Instead of combining integers, we're taking them away. Terms such as "take away" and "**minus**" are commonly used with subtraction. The answer to a subtraction problem is called the **difference**. Makes sense, right? The solution represents the difference between the two numbers involved in the subtraction operation.

Let's go back to our tripe example from earlier on. You've got 9 pieces of tripe until that tripe-stealin' neighbor of yours comes by and takes 7 pieces of your tripe. You're left with only 2 pieces of tripe.

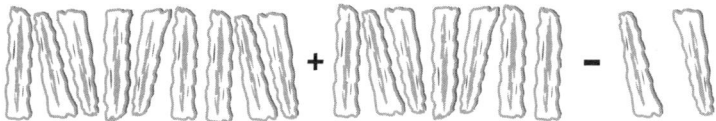

9 pieces of tripe - 7 pieces of tripe = 2 pieces of tripe

These 2 pieces of tripe are the difference between what you had before your neighbor showed up and what you have now.

As we said earlier, both the operations of addition and subtraction are best shown using the number line.

When subtracting a positive integer from any other integer, you will be moving down the number line.

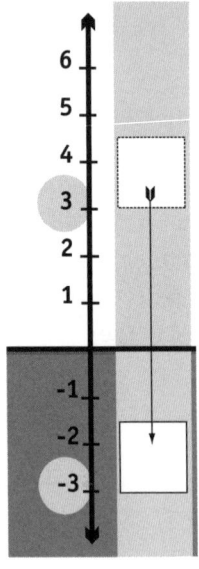

For example, $3 - 6$ means that you will start at 3 on the number line, or on the 3rd floor, and move down 6 floors. After moving down 6 floors, you are now at -3. You're in the basement again.

So we've solved another problem.

$$3 - 6 = -3$$

That's all there is to it. And remember, we said that adding a negative number is the same as subtracting by that number.

$$3 + (-6) = -3$$

$$3 - 6 = -3.$$

We get the same answer for both problems.

Now let's look at subtracting a negative integer from another negative integer.

Now, when you subtract a negative integer from any other integer, you will be moving up the number line.

That's right, even though we're subtracting, we're movin' on up. You see, when you subtract a negative integer, you end up with two negative signs in the equation, and two negative signs together in a row always equal a positive.

It's the same as addition. Just think of it this way: a double negative always equals a positive. Let's do a quick example of subtracting a negative integer.

For $3 - (-6)$, you make it read $3 + 6$.

That's because subtracting a negative number is the same as adding a positive number.

This may seem a bit confusing, but it'll all make sense in a second. We'll go back to the number line.

On the number line, you start at 3, then you move up 6 places.

Notice that after moving up 6 places you're at 9. So

$$3 - (-6) = 9$$

which is the same answer you'd get for $3 + 6$.

Remember that subtracting a negative integer is the same as adding that positive integer.

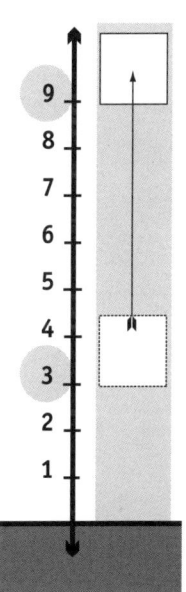

HOT TIP

If all these positive and negative rules are confusing you, then check out the **OTHER IMPORTANT STUFF** section where we show you all the rules for every type of operation.

`0:18:00`

Now let's look at subtraction with bigger numbers. In subtraction, just like addition, you generally stack one number on top of the other. You always place the amount you started with on top, and the amount being taken away on the bottom.

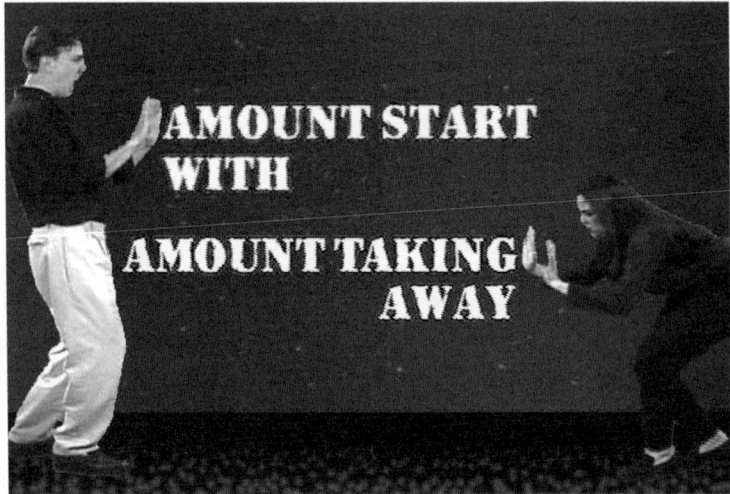

AMOUNT START WITH

AMOUNT TAKING AWAY

We subtract the bottom number from the top number in each digit place. Just like addition, we always start in the digit place

to the far right. Sometimes you may find that you have to subtract a larger number from a smaller number in a digit place. Look at this subtraction problem.

$$\begin{array}{r} 13 \\ -\ 7 \\ \hline \end{array}$$

The 7 must be subtracted from the 3 in the units place. But 7 is larger than 3. When this happens, you have to borrow.

`0:19:00`

Borrowing is done whenever the digit you are subtracting is bigger than the digit you are subtracting from.

Borrowing Rules:

✓ You must take 1 away from the digit place on top and to the left of the digit place you are subtracting

✓ Then add 10 to the number being subtracted from

✓ So anytime you borrow 1 from a digit place, then it's worth 10 to the digit place to its immediate right.

Think of it this way: if the digit you're subtracting from on top doesn't have enough, then it has to move to the place to the left to borrow 10 from that digit.

Now let's do an example where we will have to do some borrowing. We'll go back to our ant colony where a catastrophe has occurred: a gang of roaches has attacked the ant colony, enslaving part of its population. Two hundred, eighty-seven are missing from the original population of seven hundred, twenty-

three ants in the colony. We'll subtract the 287 that are missing from the original 723 to find out how many ants survived the attack.

$$723$$
$$- \underline{287}$$

A Suggestion From the Standard Deviants:

If your neighbor comes over asking you to return an item you borrowed, insist to the point of fisticuffs that you didn't borrow the item. When you wake up from your coma, laugh with your friends at your lighthearted tomfoolery.

We start subtracting on the far right. The digit place on the far right is the units place, which has a 3 and 7. So we subtract.

$$3 - 7$$

7 is larger than 3, so the 3 has to borrow from the tens place. We take away one from the tens place, which turns the 2 into a 1 now. It's always a good idea to cross out the old number and put the new number above it to make sure you don't forget that you've subtracted from the original number.

$$7\overset{1}{\cancel{2}}3$$
$$- \underline{287}$$

Remember that the 1 borrowed from the tens place is worth 10 to the units place. So we add the 10 we borrowed from the tens place to the 3 already in the units place.

$$10 + 3 = 13$$

13 is the new top number we're subtracting from.

$$\begin{array}{r} \overset{1\,13}{7\cancel{2}\cancel{3}} \\ -\ 287 \end{array}$$

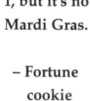

We subtract the new numbers in the units place.

$$13 - 7 = 6$$

The 6 goes in the units place of the solution.

$$\begin{array}{r} \overset{1\,13}{7\cancel{2}3} \\ -\ 287 \\ \hline 6 \end{array}$$

We've got one of the digit places done. Now we'll subtract the numbers in the next digit place on the left. We have to subtract the numbers in the tens place. Remember we borrowed, or subtracted, 1 from the 2, so in the tens place we have $1 - 8$. Again, we have to borrow from the digit place to the left in order to have a positive result for the tens place in our answer.

$$7 - 1 = 6$$

The top digit in the hundreds place now has a six in it.

$$\begin{array}{r} \overset{6\,1\,13}{7\cancel{2}\cancel{3}} \\ -287 \\ \hline 6 \end{array}$$

Subtracting 1 from the hundreds place is worth 10 to the tens place, so we add 10 to the 1 that's left.

$$10 + 1 = 11$$

Since the tens place is now $11 - 8$, we subtract these digits.

$$11 - 8 = 3$$

That gives us 3 for the tens place in the solution.

$$
\begin{array}{r}
{\scriptstyle 6\ 11\ 13} \\
7\cancel{2}\cancel{3} \\
-\ 287 \\
\hline
36
\end{array}
$$

The last place we need to deal with is the hundreds place. Since we borrowed 1 from the 7, the digits in the hundreds places are 6 (top) and 2 (bottom).

$$6 - 2 = 4$$

We put the 4 into the hundreds place of our answer.

$$
\begin{array}{r}
{\scriptstyle 6\ 11\ 13} \\
7\cancel{2}\cancel{3} \\
-\ 287 \\
\hline
436
\end{array}
$$

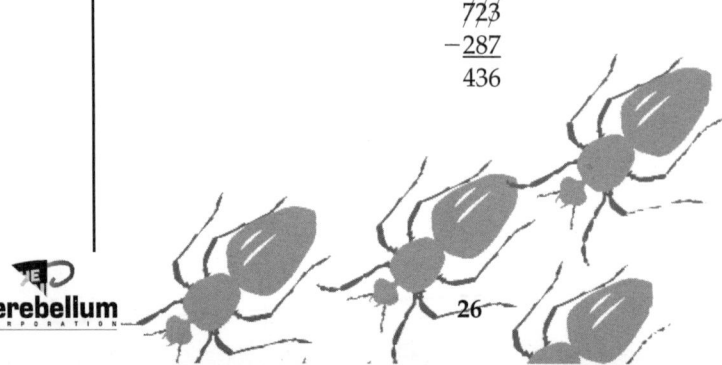

As you can see, the answer to 723 ants minus 287 ants is 436 ants. Now you know that 436 ants survived the roach attack. Now we'll move onto a third type of operation with integers.

`0:22:12`

Section C: Multiplication

Multiplication is simply an extension of addition. For instance, check out this addition problem.

$$5 + 5 + 5 = 15$$

Multiplication simplifies this problem. Instead of adding 5 + 5 + 5, you'd multiply 5 times the number of fives you were adding.

$$5 \times 3 = 15$$

As you can tell, multiplication really comes in handy when you multiply larger numbers. Multiplication saves time and effort.

By helping to save time and effort, Mathos is bringing America closer to defeating the communist threat.

The answer to any multiplication problem is called the **product**. The product of multiplying small integers, those from 12 to 0, can be found in multiplication tables. If you're not sure you remember them all, then you need to do a quick refresher and memorize them.

When you're multiplying positive and negative numbers, there are a few rules to follow.

`0:23:30`

RULES

✖ **If the numbers you're multiplying are both positive, then your product will always be positive.**

✖ **Anytime you multiply a negative number and a positive number, you will have a negative product.**

✖ **The product of multiplying two negative numbers is always positive.**

These rules are the same for any type of numbers you multiply.

Just like addition and subtraction, large numbers being multiplied are usually put one on top of the other. Arranged this way, numbers are multiplied digit by digit, just like numbers being added or subtracted.

`0:24:54`

To multiply two large numbers, you must first multiply the digit at the far right of the bottom number by every digit of the top number, one-by-one, right to left, carrying over if necessary.

Other
Records

Joe
Braband
389
somersaults

DJ
Brown
450
somersaults

Phil
Stuben
200
somersaults

Rhode
Island

Cerebellum Sports Update

Deviant
TV CH. 2

Let's look at a normal multiplication operation. Ronald's attempting to become the first person to somersault across a whole state. An amazing achievement, even if it is Rhode Island.

He's been somersaulting for 23 days now. He somersaults 186 times a day. How many somersaults has he done in all? To figure this out, we multiply.

$$\begin{array}{r} 186 \\ \times\ 23 \\ \hline \end{array}$$

You always start the multiplication process by multiplying the bottom number on the far right by every digit in the top number. For this problem, the units place is on the far right.

The 3 is in the units place of the bottom number. The first thing we do to solve this problem is to multiply 3 by all the digits in the top number.

$$3 \times 6 = 18$$

REMEMBER that for any math problem, whether you're multiplying, adding, subtracting, or dividing, you always want to keep the digit places lined up.

We can only have one digit for each digit place, so we leave the 8 and carry the 1 to the tens place. No prob.

$$
\begin{array}{r}
{}^{1} \\
186 \\
\times\ 23 \\
\hline
8
\end{array}
$$

Now we multiply the 3 by the next number to the left in the top of our multiplication problem.

$$3 \times 8 = 24$$

We must also add the 1 we carried over, which leaves us with

$$24 + 1 = 25$$

Your meal ticket to good grades, the respect of your peers, and backstage passes to the Commodores.

– Albert Einstein, commenting on *The Zany World of Basic Math*

The 5 stays in the tens place of the result, and we carry the 2 to the digit place to the left.

$$\begin{array}{r} {\scriptstyle 2\,1} \\ 186 \\ \times\ \underline{23} \\ 58 \end{array}$$

Now we just multiply 3 by the last digit on top and add the 2 that we carried over from the tens place product.

$$3 \times 1 = 3$$

$$3 + 2 = 5$$

We leave the 5 in the hundreds place.

$$\begin{array}{r} {\scriptstyle 2\,1} \\ 186 \\ \times\ \underline{23} \\ 558 \end{array}$$

All right, we're halfway through the problem. All we've got to do is follow the same process we just did for the number in the units place for each digit in the bottom number. But first we have to take one step before multiplying the tens-place digit in the bottom number.

Notice that the 2 is in the tens place. When we multiply the 2 by the number in the units place on top, we have to put the product of those two numbers in the tens place of the result. But that means there will not be a number in the units place of our second result.

Why in the world are we putting the product in the tens place and not the units place? Well, notice that the results are placed one under the other.

$$
\begin{array}{r}
{\scriptstyle 2\,1} \\
186 \\
\times\ \underline{23} \\
558\ \text{(Result 1)} \\
\text{(Result 2)}
\end{array}
$$

The first result is the product of the digit in the units place of the bottom number (3), times the top number in our multiplication problem (186). The second result is the product of the bottom digit in the tens place, 2, times the top number in our multiplication problem. Now, listen closely to this next bit of information.

FLASH!

When you're multiplying the bottom digit times the top number, the units place of your result should always go in the *same digit place* as the bottom digit you're multiplying by. Place zeroes in any unused spaces to the right of that product's units place.

What this means for us right now: We put a 0 underneath the 8 in the units place of our second result. That's because the bottom digit we're multiplying by, 2, is in the tens place, and we want the units place of the product of 2 times the top number to go in the tens place as well.

The digit at the far right of the top number is 6. When we multiply 2 times 6, their product will go underneath the 2 in the tens place of the second result. So we put a 0 underneath the 8, that is, in the units place of our second result.

$$
\begin{array}{r}
186 \\
\times\ 23 \\
\hline
558 \\
0
\end{array}
$$

Now we just multiply all the digits on top by 2.

$$2 \times 6 = 12$$

The 2 goes in the tens place of the second result. We carry the 1 to the next column.

$$
\begin{array}{r}
1 \\
186 \\
\times\ 23 \\
\hline
558 \\
20
\end{array}
$$

Next we multiply the 2 by the next digit to the left and add the carried 1.

$$2 \times 8 = 16$$

$$16 + 1 = 17$$

The 7 stays in the hundreds place of the second result and the 1 is now carried over.

$$\begin{array}{r} {}^{1\,1} \\ 186 \\ \times\,23 \\ \hline 558 \\ 720 \end{array}$$

We have one more multiplication operation left.

$$2 \times 1 = 2$$

We add the carried 1.

$$2 + 1 = 3$$

We put the 3 in the thousands place of the second result.

$$\begin{array}{r} {}^{1\,1} \\ 186 \\ \times\,23 \\ \hline 558 \\ 3720 \end{array}$$

After you've multiplied all the bottom numbers by the top numbers, and put the zeroes in the appropriate digit places, you can add up all your results.

For this example, we only have 2 results because the bottom number only has 2 digit places. If we'd had 3 digits in the bottom number, then we would have had 3 results to add together.

QUICK TIP

There are always the same number of results as there are digits in the bottom number.

The last thing we need to do is add the results.

$$
\begin{array}{r}
558 \\
+\ \underline{3720} \\
4{,}278
\end{array}
$$

So that's the product and the answer to our multiplication problem.

$$
\begin{array}{r}
186 \\
\times\ \underline{23} \\
558 \\
\underline{3720} \\
4{,}278
\end{array}
$$

That's how many somersaults Ronald did in all. Next we'll look an extension of multiplication: exponents.

Harry Stotomeitz, the man who proved multiplication of positive and negative numbers.

`0:30:44`

Section D: Exponents

Exponents are a shorthand notation for the multiplication of a number times itself. What do we mean by that? Look at this example.

$$5 \times 5 \times 5 = 125$$

$$5^3 = 125$$

Both of these expressions represent the same amount. The 3 is known as the exponent, and the 5 is called the base. The exponent shows how many times the base number is multiplied times itself.

OTHER COMMON EXAMPLES OF BASES

A MILITARY
BASE

A BASE
(AS IN "ACID & BASE"

FIRST BASE?

FIRST BASE

Here's another example of an exponent:

$$3^4 = 3 \times 3 \times 3 \times 3 = 81$$

In the following example, we have a negative number for our base, but that doesn't change the way we find our answer. Notice the answer is negative. That's because we multiplied a negative base an odd number of times.

$$(-5)^3 = -5 \times -5 \times -5 = -125$$

Now, there are a few exponents which do the same thing to every base number no matter what the base number may be.

THE POWER OF ONE

✔ Any integer raised to the first power is the same as that integer.

$$8^1 = 8$$

$$-15^1 = -15$$

There's one other exponent that it's really important for you to remember.

✔ Any non-zero integer raised to the zero power equals one. Like these:

$$1500^0 = 1$$

$$2^0 = 1$$

One more thing about exponents that you can tuck away in your memory bank. Look at these numbers. Each time we add a zero, we increase by one digit place.

10,000

1,000

100

10

1

Now look at these exponents. Each time we increase the exponent by one, we increase the number of digit places in the corresponding number by one.

$$10,000 = 10^4$$
$$1,000 = 10^3$$
$$100 = 10^2$$
$$10 = 10^1$$
$$1 = 10^0$$

This is just a quick note to show you how each digit place in a number increases by a power of ten. Also notice that the amount of the exponent equals the number of zeroes in each number. Exponents are going to be extra helpful later on when you start dealing with extremely large numbers.

Section E: Division

`0:32:59`

Just like subtraction is the reverse of addition, division is the reverse of multiplication. We'll look at the division of integers now and deal with other forms of division later in the tape. **Division** is the process of breaking down numbers into equal groups.

There are 3 parts to every division problem.

`0:33:52`

✔ The number being divided = **dividend**

✔ The number we're dividing by = **divisor**

✔ The answer to a division problem = **quotient**

For example, Big Sam has 27 people he wants to beat up. But he's not all bad; he's a man who believes in sharing. He's going to divide these soon-to-be black-and-blue people equally between his 3 friends, which means that he's dividing 27 by 3.

$$27 \div 3 = 9$$

In our example, 27 is the number being divided, so it's the **dividend**. The number we're dividing by is the 3 so it's the **divisor**. 9 is the answer to our division problem so it's the **quotient**.

Remember we said that division is multiplication in reverse. To show you what we mean, let's reverse the division process on our last example. We'll multiply the quotient by the divisor. If their product is the dividend, then we did the division problem right.

$$9 \times 3 = 27$$

Our answer is the dividend, so we know we did our division problem right. Pretty cool, huh?

So division is just multiplication in reverse. Just like everything else, the division process becomes more involved when we're dividing larger numbers. Picture this problem.

$$581 \div 34 = ?$$

We're going to find out how many thirty-fours are in five hundred, eighty-one. First, we'll put the numbers into the proper format. The dividend, 581, goes underneath the division sign. It's being divided. The divisor, 34, goes to the left of the division sign.

$$34\overline{)581}$$

Dividing large numbers like these is called **long division**. The first step is to see how many times the divisor, 34, will go into the number in the largest digits place in the dividend. The largest digit place will always be the number on the far left, since digit places increase to the left. In this instance, the largest digits place is the 5 in the hundreds place. So we divide 34 into 5. We've hit a stumbling block. 5 can't be divided by 34 to get a whole number.

$$5 \div 34$$

When this happens, you must include the digit to the right. So we have to include the integer in the digits place to the right, 8. Now we see how many times 34 will go into 58, and we discover that 34 goes into 58 only 1 time.

$$34\overline{)581}^{1}$$

If 58 could not have been divided by 34, then we would have had to include the next digit to the right. Our problem looks like this now.

$$34\overline{)581}^{1}$$

Notice the 1 goes above the 8 and not the 5. That's because 34 is divided into 58, and not just into 5.

The next step is to multiply 1 by 34.

$$1 \times 34 = 34$$

You place the product of these two numbers directly below 58.

$$
\begin{array}{r}
1 \\
34\overline{)581} \\
34
\end{array}
$$

Now we subtract 34 from 58.

$$
\begin{array}{r}
1 \\
34\overline{)581} \\
\underline{34} \\
24
\end{array}
$$

Before we divide 34 into this number, we first bring down the 1 in the units place of the dividend next to 24.

$$
\begin{array}{r}
1 \\
34\overline{)581} \\
\underline{34} \\
241
\end{array}
$$

Now we're trying to find out how many times 34 goes into 241. The only way to find out the answer to this is through trial and error. Let's try to divide 34 into 241 six times. To do this, we multiply 6 × 34.

$$
\begin{array}{r}
34 \\
\times\ 6 \\
\hline
204
\end{array}
$$

We subtract this amount from 241.

$$
\begin{array}{r}
241 \\
-\ 204 \\
\hline
37
\end{array}
$$

44

We have 37 left over, but the divisor, 34, can be divided into that number because 37 is larger than 34. We know we can divide 34 into 241 by a larger number than 6. Let's try 7.

$$\begin{array}{r} 34 \\ \times\ 7 \\ \hline 238 \end{array}$$

238 goes beneath 241 because we're dividing 34 into 241.

$$\begin{array}{r} 17 \\ 34\overline{)581} \\ \underline{34} \\ 241 \\ 238 \end{array}$$

Now we subtract 238 from 241.

$$\begin{array}{r} 17 \\ 34\overline{)581} \\ \underline{34} \\ 241 \\ \underline{238} \\ 3 \end{array}$$

3 can't by divided by 34 so we know that 34 goes into 241 seven times.

$$\begin{array}{r} 17 \\ 34\overline{)581} \\ \underline{34} \\ 241 \\ \underline{238} \\ 3 \end{array}$$

There are no more numbers to bring down from the dividend, so we're done because 34 can't go into 3. But we still have a remainder of 3 left at the end of the division problem.

`0:37:45`

A division problem has a **remainder** when the divisor cannot go into the dividend exactly. The remainder should always be less than the divisor. If it's not, then you know you've done something wrong.

So the answer to our division problem is 17 with a remainder of 3. It's also always a good idea to check your work to make sure some minor error hasn't thrown off the whole division process.

`0:38:15`

To check our answer, the first step is to multiply 34 by 17.

$$\begin{array}{r} 34 \\ \times\ 17 \\ \hline 578 \end{array}$$

Now we just add the remainder to our product.

$$\begin{array}{r} 578 \\ +\ \ 3 \\ \hline 581 \end{array}$$

581 is our dividend, so our answer is right.

REMEMBER: WHEN CHECKING...

✓ To check a division problem, you just multiply the quotient, or answer, by the divisor and add the remainder (if there is one). If your result equals the dividend, then you know you're right.

✓ To check a hockey player, you always want to focus your eyes on the shoulders of the man you plan to check. Aim a little ahead of him, to account for his movement. Take no more than three strides in preparing your check, or else you'll be charging, which is illegal. Right before colliding with your man, lift your arms a bit and use all your available strength to propel your man into the boards. When he collapses to the ice, skate quickly away, so as to avoid vicious reprisals.

Section F: Order of Operations

`0:39:11`

Standard
Order of Operations

In many math problems, you'll have a long string of numbers separated by different combinations of operators. You might have to multiply, divide, and add, in addition to dealing with parentheses and exponents all in the same problem. Here are some examples.

$$5^2 + (9 - 3) \times 4 - 8 \div 2 = ?$$

$$-8 - \frac{9}{-3} \times (-1 + 6) + 7^2 = ?$$

Don't worry, a standard **order of operations** has been developed just for these types of problems. You have to do certain kinds of operations before others, and you always work from left to right with these strings of numbers.

48

Lots of people use a mnemonic device to remember this order of operations. The first letter of each word in the order of operations spells **PEMDAS** (**P**arentheses, **E**xponents, **M**ultiplication, **D**ivision, **A**ddition, **S**ubtraction. Pronounce it just like it's spelled: "Pem-das").

REMEMBER, the standard order of operations is parentheses first, followed by exponents, then multiplication and division, and addition and subtraction operations are last.

Now, you might be thinking, PEMDAS? What the hell does that spell? Well, nothing, actually, which kinda limits its usefulness as a mnemonic. However, we're hoping that if you read PEM-DAS over and over again, then you'll remember it anyway.

PEMDAS, PEMDAS,

PEMDAS, PEMDAS, PEMDAS, PEMDAS, PEMDAS, PEMDAS,
PEMDAS, PEMDAS, PEMDAS, PEMDAS, PEMDAS, PEMDAS,
PEMDAS, PEMDAS, PEMDAS, PEMDAS, PEMDAS, PEMDAS,
PEMDAS, PEMDAS, PEMDAS, PEMDAS, PEMDAS, PEMDAS,
PEMDAS, PEMDAS, PEMDAS, PEMDAS, PEMDAS, PEMDAS,
PEMDAS, PEMDAS, PEMDAS, PEMDAS, PEMDAS, PEMDAS,
PEMDAS, PEMDAS, PEMDAS, PEMDAS, PEMDAS, PEMDAS,
PEMDAS, PEMDAS, PEMDAS, PEMDAS, PEMDAS, PEMDAS,
PEMDAS, PEMDAS, PEMDAS, PEMDAS, PEMDAS, PEMDAS,
PEMDAS, PEMDAS, PEMDAS, PEMDAS, PEMDAS, PEMDAS,
PEMDAS, PEMDAS, PEMDAS, PEMDAS, PEMDAS, PEMDAS,
PEMDAS, PEMDAS, PEMDAS, PEMDAS, PEMDAS, PEMDAS,
PEMDAS, PEMDAS, PEMDAS, PEMDAS, PEMDAS, PEMDAS,
PEMDAS, PEMDAS, PEMDAS, PEMDAS, PEMDAS, PEMDAS,
PEMDAS, PEMDAS, PEMDAS, PEMDAS, PEMDAS, PEMDAS,
PEMDAS, PEMDAS, PEMDAS, PEMDAS, PEMDAS, PEMDAS,
PEMDAS, PEMDAS, PEMDAS, PEMDAS, PEMDAS, PEMDAS,
PEMDAS, PEMDAS. PEMDAS, PEMDAS, PEMDAS, PEMDAS,
PEMDAS, PEMDAS, PEMDAS, PEMDAS, PEMDAS, PEMDAS,
PEMDAS, PEMDAS, PEMDAS, PEMDAS, PEMDAS, PEMDAS,
PEMDAS, PEMDAS, PEMDAS, PEMDAS, PEMDAS, PEMDAS,
PEMDAS, PEMDAS, PEMDAS, PEMDAS, PEMDAS, PEMDAS,
PEMDAS, PEMDAS, PEMDAS, PEMDAS, PEMDAS, PEMDAS,
PEMDAS, PEMDAS, PEMDAS, PEMDAS, PEMDAS, PEMDAS,
PEMDAS, PEMDAS, PEMDAS, PEMDAS, PEMDAS, PEMDAS,
PEMDAS, PEMDAS, PEMDAS, PEMDAS, PEMDAS, PEMDAS,
PEMDAS, PEMDAS, PEMDAS, PEMDAS, PEMDAS, PEMDAS,
PEMDAS, PEMDAS, PEMDAS, PEMDAS, PEMDAS, PEMDAS,
PEMDAS, PEMDAS, PEMDAS, PEMDAS, PEMDAS, PEMDAS,
PEMDAS, PEMDAS, PEMDAS, PEMDAS, PEMDAS, PEMDAS,
PEMDAS, PEMDAS.PEMDAS, PEMDAS, PEMDAS, PEMDAS,
PEMDAS, PEMDAS, PEMDAS, PEMDAS, PEMDAS, PEMDAS.

SINKING IN, ISN'T IT?

Here's an example of a math problem that you must figure out using the order of operations to get the right answer.

$$5^2 + (9 - 3) \times 4 - 8 \div 2 = ?$$

Now let's solve this puppy.

Moving from left to right, we have to do any operations in parentheses first.

PEMDAS

$$9 - 3 = 6$$

So we substitute 6 for the $9 - 3$ in parentheses.

HOT TIP

If the string of integers contains any grouping symbols, like parentheses, brackets, or curly brackets, then perform the operations inside the grouping symbols FIRST. If you see one set of grouping symbols INSIDE another set, do every operation inside the innermost group, and then do every operation inside the outermost group. **Remember to follow the order of operations for the operations inside the grouping symbols.**

The problem now looks like this.

$$5^2 + 6 \times 4 - 8 \div 2 = ?$$

Since there are no more parentheses left, we move onto the next step in the order of operations, exponents. Again moving from left to right, we find 5 to the 2nd power.

Remember, the **exponent** is just the number of times you multiply the base number by itself.

P**E**MDAS

$$5^2 = 5 \times 5 = 25$$

We put 25 in place of 5^2. Our problem now reads

$$25 + 6 \times 4 - 8 \div 2 = ?$$

There are no more exponents. Multiplication and division are the next step in the order of operations. Starting on the left, the first multiplication problem we find is 6×4. So we multiply this out and replace the multiplication operation with the product.

PE**M**DAS

$$6 \times 4 = 24$$

$$25 + 24 - 8 \div 2 = ?$$

We continue moving to the right and find a division problem.

PEM**D**AS

$$8 \div 2 = 4$$

The 4 goes into our problem and replaces the division operation.

$$25 + 24 - 4 = ?$$

Addition and subtraction operations are all that's left.

PEMD**AS**

Beginning on the left, we add

$$25 + 24 = 49$$

49 replaces the addition operation, leaving us just one subtraction problem.

$$49 - 4 = 45$$

Sooooo, that means that the answer to five to the second power, plus nine minus three, times four, minus eight divided by two, is forty-five. Okay. Everyone take a breather. Let's procrastinate!

I'm not saying I'm a freakin' lunatic, but if that otter in the corner doesn't stop throwin' stuff at me, I'm gonna get really mad.

–Kristie W.

PROCRASTINATION SUGGESTIONS

Here are a few suggestions of what to do when you don't feel like doing math:

1. Do TV Guide Crossword puzzle.
2. Build card condos.
3. Build makeshift boxing ring in kitchen.
4. Look at recently expired milk and see if it's safe to drink.
5. Try to organize spiders on ceiling into battalions.
6. Try to organize spiders into labor unions.
7. Clean out cup with toothbrushes in it.
8. Bake a cake.
9. Go to sleep.

If you're going to use a calculator, then it's important to see if your calculator has a specific memory set-up for the order of operations. To do a complex string of arithmetic on a calculator, you must either have a calculator that recognizes the order of operations, or you have to break up the string yourself and only use the calculator to figure out specific operations within the string. In that case, you still need to use the order of operations. (PEMDAS, PEMDAS, PEMDAS. . . .)

Section G: Rounding Off

All right, let's say you've got two big numbers to add but you don't have time to get your paper out and do a bunch of carrying and so on. Well, then you can round those numbers off. **Rounding off** is a method used when you're not working with exact numbers. Whenever you round off, you do it by digit place. You can round off to the tens place, the hundreds place, the thousands place, and so on and so forth, until infinity, depending on your needs.

Let's do an example. We'll round this number off to the nearest hundred.

Anytime you want to round off a positive integer, you have to take 4 easy-to-follow steps.

C'mon! How
many Indians
could there
possibly be?

– General
Custer

Step One: Find the digit place you're rounding off to.
For our number, the digit in the hundreds place is a 4.

Step Two: Look one digit place to the right of where you
want to round off to.

If that digit is less than 5 → go to step number 4.

If that digit is greater than or equal to 5 → go to
step number 3.

Step Three: Take the digit in the digit place you are rounding, and increase it by one. We're trying to round off to
the nearest hundreds place, so the digit place immediately
to the right is the tens place. And it has a 9 in it. Since 9
is greater than 5, we increase the 4 in the hundreds place
by one. If the number in the tens place had been less
than 5, then we would have left the 4. There's still
one more step left.

$$1492 \rightarrow 1592$$

Step Four: Use zeroes to replace all digits to the right of the
digit place you're rounding off to.

Since we are rounding off to the nearest hundreds place, we
replace everything to the right of the 5 with zeroes.

$$1592 \rightarrow 1500$$

We have our answer. But what if we'd wanted to round 1,492 to a different digit place, like, say, the nearest thousands place? Well, the number would be 1000. That's because the number to the right of the thousands place is a 4. It's less than 5, so we skip to step four and replace all numbers to the right of the 1 with zeroes.

$$1492 \rightarrow 1000$$

You can round off any integer, no matter how large it is, using the steps we've just talked about.

57

`0:46:20`

Section H: Scientific Notation

Another method used to make it easier to calculate with large numbers is **scientific notation**.

Friendly Reminder: we said earlier that each digit place of an integer is represented by a power of ten

$$10^1 = 10$$
$$10^2 = 100$$

and so on.

> **REMEMBER**, in scientific notation, you multiply a number between 1 and 10 by one of these powers of ten.

Let's do an example. We'll express 2000 in scientific notation. The number 2000 becomes 2×10^3 in scientific notation. How do we know this? Well, there are 3 zeroes in 2000. And, like we said earlier, as long as your left-most digit is the only number different from zero, the amount of your exponent will always be equal to the amount of zeroes in your number. This means that 3 is our power of ten here.

$$10^3 = 1000$$

Now we just multiply 1000 by the first digit in the number we're trying to express. In this case, the first digit is a 2.

$$
\begin{array}{r}
1000 \\
\times \quad 2 \\
\hline
2000
\end{array}
$$

And presto–

$$2 \times 10^3 = 2000$$

We know all of you out there want to do another. Quick!

We're going to put 400,000 into scientific notation. The first thing we need to do is to count the number of zeroes to determine what power of ten we're working with. There are 5 zeroes in 400,000. Since there are 5 zeroes, you know that 5 is your exponent.

$$5 \text{ zeroes} \rightarrow 10^5$$

We're halfway there. We know our power of ten is 5, or 100,000. Now that we have the right number of digit places, all we need to do is to multiply 100,000 times 4, the first digit in the number we're trying to express.

$$\begin{array}{r} 100{,}000 \\ \times \underline{4} \\ 400{,}000 \end{array}$$

Aaand we're done. That was a quickie.

$$4 \times 10^5 = 400{,}000$$

Don't forget that a large number is in scientific notation when it is written as a number between 1 and 10 multiplied by a power of ten. Look at the expression 4×10^5. The 4 is between 1 and 10.

We've only been looking at numbers whose left-most digit is different from zero. After you study decimals, you'll be able to use scientific notation for numbers with 2 digits different from zero at the beginning of the number, like 162,000 or 29,000,000. So let's get to those no good stinkin' decimals. . .right after you take this quiz.

Quiz 1

1. Add the following integers. <u>whole number</u>

    a. 35 + 6

    b. 245 + 397

2. Subtract the following integers.

    a. 35 − 6

    b. 100 − 38

3. Multiply the following integers.

    a. 41 × 27

    b. 135 × 89

4. Simplify the following by removing the exponents.

    a. 2^4

    b. 16^2

5. Divide the following integers.

    a. 512 ÷ 8

    b. 1690 ÷ 13

6. Solve the following strings of integers, paying careful attention to the order of operations.

    a. $2 + 3 \div 1 - 2$

    b. $2 \times (14 - 9) - 1$

7. Round the following integers off to the nearest hundred.

    a. 1,545

    b. 291

8. Write the following integers in scientific notation.

    a. 1,000,000

    b. 3,000

Decimals

`0:49:53`

In the previous section, we explored integers. We looked at the operations of addition, subtraction, multiplication, and division. The next type of number we'll look at is decimals. Decimals represent parts of whole numbers and can be either positive or negative.

Section A: The Decimal Point and Decimal Places

`0:50:31`

We've said several times so far that each digit place represents multiplication by a power of ten. The same is true to the right of the decimal point. The decimal point is to the right of the units place. Decimals are actually in every number you see but are not written with integers. That is, the number 42 could be written as 42.0, but it's a little cumbersome to do this.

REMEMBER, the decimal point represents the separation of the units place and the tenths place.

The place to the right of the tenths place is the **hundredths place**. To the right of the hundredths place is the **thousandths place** and so forth. It's the same as to the left of the decimal, except you just add a "ths" to the end of the place name. For example, 52.3 is read as "fifty-two and three tenths." Or, 7.69 is read as "seven and sixty-nine hundredths."

A Standard Deviant Suggestion:

Become the life of the next party you're at by adding a "ths" to the end of everyone's name each time you say it. When you wake up from your coma, laugh with your friends at your light-hearted tomfoolery.

Addition, subtraction, multiplication and division with decimals are a lot like addition, subtraction, multiplication, and division with integers. The important difference is that you have to keep track of the decimal places you're working with.

Section B: Adding and Subtracting with Decimals

`0:52:28`

When you're adding or subtracting decimals, the decimal points in the numbers you're adding or subtracting must line up. For example, if you want to add 56.3 and 7.89, then each number's decimal points must line up. For this problem, we line up the 6 and the 7 because they are both in the units place of their respective numbers.

Yes, let them eat cake. See what they say.

– Marie Antoinette

$$
\begin{array}{r}
56.3 \\
+\ \underline{7.89}
\end{array}
$$

Now that the numbers are lined up, we can add them just like we did with integers, carrying when needed. To clarify our addition problem, we can place a zero to the right of the 3 without changing the value of the first number, since it's to the right of the decimal. What do we mean by that? Well, look at these two numbers.

$$56.3000000000$$

$$56000000000.03$$

In the first number, the zeroes could be dropped without changing the value of the number. It would still be 56.3 after we dropped the zeroes. For the second number, this is not possible. The second number is fifty-six billion point zero three. If we

65

drop the zeroes, it would be fifty-six point three. All the zeroes are important; they all affect the size of the number. Now that we've put in a zero, let's go back and add.

$$56.30$$
$$+\ 7.89$$

Our first step in the problem is to add the numbers to the farthest right in the hundredths place.

$$9 + 0 = 9$$

We put the 9 into our sum just like when we were adding integers.

$$56.30$$
$$+\ 7.89$$
$$9$$

We move to the left and add the digits in the next digit place.

$$8 + 3 = 11$$

We put a 1 in the tenths place of the sum and carry the 1 in the tenths place of 11 to the units place of our problem.

$$\overset{1}{}$$
$$56.30$$
$$+\ 7.89$$
$$19$$

As we said earlier, it's important to keep the decimal points lined up. So we just bring down the decimal point to line up exactly with the decimal points in both of our numbers.

$$
\begin{array}{r}
1 \\
56.30 \\
+\ \underline{7.89} \\
.19
\end{array}
$$

Now we finish up by adding the rest of the numbers.

$$7 + 6 + 1 = 14$$

The 4 stays in the units place, and we carry the 1 to the tens place of our addition problem.

$$
\begin{array}{r}
11 \\
56.30 \\
+\ \underline{7.89} \\
4.19
\end{array}
$$

Now we add the numbers in the tens place.

$$1 + 5 = 6$$

The 6 goes in the tens place of the solution and presto!

$$
\begin{array}{r}
11 \\
56.30 \\
+\ \underline{7.89} \\
64.19
\end{array}
$$

67

The answer to our addition problem is sixty-four point nineteen, or sixty-four and nineteen hundredths.

Subtracting with decimals is done in the same way as adding. The most important thing to remember is to line the decimal points up.

AS PRESIDENT OF THE CENTER FOR THE ADVOCACY OF CUMBERSOME-NESS AND **EXCESSITUDE**, I'D LIKE TO SPEAK ON BEHALF OF CUMBER-SOME METHODS..

PRESIDENT OF T.C.F.T.A.O.C.A.E.

LISTEN, BUDDY, YOU'RE WASTING MY TIME.

THESE PEOPLE GOT MATH TO LE WE'RE GONNA D THE **SIMPLE** WAY.

THANK YOU.

CENTER FOR SIMPLICITY

Section C: Multiplying and Dividing Decimals

`0:54:47`

For multiplication and division of numbers, you must also keep track of the decimal places. If you don't, they'll muck the whole thing up.

Multiplication of decimals is done exactly the same as multiplication of integers:

- One number is set on top of the other.
- Each digit of the bottom number is multiplied by all digits in the top number.
- The appropriate zeroes are put into each result.

The results are then added together, just like we did with integers. But we need to take one additional step when multiplying decimals.

REMEMBER, you have to count the total number of digit places to the right of the decimal point in the two numbers that are being multiplied. Then you move the decimal point THAT number of decimal places to the left in the product.

So if there are a total of 4 numbers to the right of the units place in the numbers being multiplied, then the decimal is moved 4 digit places to the left from the right end of the product. Let's do an example.

$$1.86$$
$$\times \underline{\ 2.3\ }$$

We figure out the answer to this problem the same way that we would for 186 times 23. We go through the normal steps in calculating a multiplication problem.

$$186$$
$$\times \underline{\ 23\ }$$
$$558$$
$$\underline{3720}$$
$$4278$$

Now we need to adjust the decimal point in our product. There are 3 digits to the right of the decimal points in both of the numbers we multiplied. So we have to insert a decimal point 3 decimal places from the right end of our product. That means we stick the decimal point in between the 4 and 2.

$$1.86$$
$$\times \underline{\ 2.3\ }$$
$$558$$
$$\underline{3720}$$
$$4.278$$

The answer is (drum roll) four and two hundred seventy-eight thousandths.

We divide decimals exactly the same way we divide integers, except that we move the decimal points BEFORE dividing.

Here's an example.

$$1.45 \div .5 = ?$$

1.45 goes inside the division sign, since it's the dividend, or number being divided. 0.5 goes to the left of the division sign.

$$.5\overline{)1.45}$$

We take 2 steps before dividing.

> The first step any time you're dividing decimals is to move the decimal point in the divisor out of the way, all the way to the right. Place it so that the last digit of the divisor is in the units place.

For our example, the decimal point in the divisor (0.5) must be moved 1 digit place to the right. That way, the last number in the divisor is in the units place.

$$0.5 \rightarrow 5.0$$

$$5\overline{)1.45}$$

Okay, we've got to take one more step before dividing.

> The second step is to move the decimal point in the dividend the SAME number of places to the right as we moved the divisor's decimal point.

So we move the decimal point in the dividend (1.45) the same number of places to the right as we moved the divisor's

decimal point. We moved the divisor's decimal point 1 place, so now we move the decimal in the dividend 1 place.

You make my
spider sense
tingle.

–Spiderman

$$1.45 \rightarrow 14.5$$

$$5\overline{)14.5}$$

Our division problem is now $14.5 \div 5$. No sweat, right? Now we just go through the normal division process. We divide 5 into 14.

$$5\overline{)14.5}^{2}$$

We multiply 2 by our divisor and put that product beneath 14.

$$2 \times 5 = 10$$

$$\begin{array}{r} 2 \\ 5\overline{)14.5} \\ 10 \end{array}$$

Now we subtract 10 from 14 and bring down the 5, the next digit to the right in our dividend.

$$\begin{array}{r} 2 \\ 5\overline{)14.5} \\ \underline{10} \\ 45 \end{array}$$

We divide 45 by 5.

$$\begin{array}{r} 2.9 \\ 5)\overline{14.5} \\ \underline{10} \\ 45 \end{array}$$

Then we multiply 9 times 5 and put the product beneath 45.

$$9 \times 5 = 45$$

$$\begin{array}{r} 2.9 \\ 5)\overline{14.5} \\ \underline{10} \\ 45 \\ \underline{45} \end{array}$$

5 times 9 equals 45, so we have no remainder.

$$\begin{array}{r} 2.9 \\ 5)\overline{14.5} \\ \underline{10} \\ 45 \\ \underline{45} \\ 0 \end{array}$$

Our last step is to insert a decimal point into the answer so that it's lined up with the decimal point in the dividend. The decimal place directly above the decimal point in 14.5 is between the 2 and the 9.

$$\begin{array}{r} 2.9 \\ 5\overline{)14.5} \\ \underline{10} \\ 45 \\ \underline{45} \\ 0 \end{array}$$

This leaves us with a quotient of 2.9, or two and nine tenths.

REMEMBER...

✔ You move the decimal point AFTER you multiply. You count the number of digits to the right of the decimals you're multiplying, then put the decimal that number of digit places to the left in your final result. You adjust the decimal point *before* you begin the problem in division.

✔ For division, you move the decimal in the divisor so that it's to the right of the units place. Then you move the decimal in the dividend the same number of places (also to the right).

74

Section D: Rounding Off Decimals

`0:59:21`

Rounding off decimals is similar to rounding off integers. However, when you round off numbers with decimals in them, you may be rounding off to the right of the decimal point as well. The steps for rounding off positive decimals are the same as the steps for rounding positive integers.

- ✔ You figure out which digit place you're rounding off to.

- ✔ Then, you look at the number to the right of that digit place. If it is equal to or greater than 5, then round up the digit you're rounding off to. If it's less than 5, then don't do anything!

- ✔ Your last step is to use zeroes to replace all the digits to the right of the digit place you're rounding to.

Let's do some rounding. Check this number out.

$$3.814$$

We'll round this decimal to the nearest units place. The number to the right of the units place is an 8, which is greater than 5, so we round the units-place digit up one from 3 to 4.

$$3.814 \rightarrow 4.814$$

STUDY SIDEKICK

Our last step is to replace all the digits to the right of the 4 with zeroes, but since all these digits are to the right of the decimal, and at the far right of the number, they don't affect the value of the number.

$$4.814 \rightarrow 4.000 \rightarrow 4$$

Our answer is 4.

You can see that the process of rounding off a number with a decimal point is pretty much the same thing as rounding off an integer.

Let's round some more!!

$$3.814$$

If we round this number to the nearest tenth, we get 3.8, or three and eight tenths.

$$3.814 \rightarrow 3.8$$

Rounding to the nearest hundredth gives us 3.81, or three and eighty-one hundredths.

$$3.814 \rightarrow 3.81$$

In both cases, the digit to the right of the digit being rounded off is less than 5, so the digit being rounded off does not need to be raised one. Notice that we didn't put zeroes to the right of the last digit. The value of the rounded number doesn't change.

EVERYDAY OBJECTS

the tire

a billiard ball

the doughnut

the Earth

Before They Were Rounded Off

The Tire. A Billiard Ball. The Earth. The Doughnut.

STUDY SIDEKICK

`1:01:10`

Section E: Scientific Notation with Decimals

Decimals are often used in scientific notation. Remember, we are dealing with powers of ten. All we're really doing in scientific notation is moving the decimal point enough places so that our number is between 1 and 10, then multiplying it by a power of ten. The number of places we move our decimal point tells us what power of ten we multiply by.

$$1.44,000,000,000$$

(places: 11 10 9 8 7 6 5 4 3 2 1)

$$1.44 \times 10^{11}$$

$$10^{11} = 100,000,000,000$$

$$1.44 \times 100,000,000,000$$

We can use scientific notation to show the distance from the earth to the sun, 144,000,000,000 meters.

78

First, we move the decimal place to the left so that the number remaining is greater than or equal to 1, and less than 10. For our example, we move the decimal point over to the left until it's between the 1 and the 4.

$$1.44000000000$$

Our decimal point is now 11 places to the left of where it was before. As you recall, the number of places you move the decimal point is the value of the exponent. So we know our power of ten and the number it's multiplied by.

$$1.44 \times 10^{11}$$

Notice that ten to the eleventh power is one hundred billion, and 1.44 times one hundred billion is one hundred, forty-four billion meters.

$$10^{11} = 100,000,000,000$$

$$1.44 \times 100,000,000,000$$

$$= 144,000,000,000$$

Up to this point, we've been moving the decimal point to the left, and the power of ten was the number of decimal places we moved. For numbers less than 1, we have to move the decimal to the RIGHT to make the number we multiply the power of ten by ≥ 1 and < 10. The power of ten is still represented by the

number of places we move the decimal, but in this case, the power of ten is negative since we move to the right. For example, we'll put the following decimal into scientific notation.

$$0.0003789$$

First, we must move the decimal point 4 places to the right.

$$3.789$$

Now we find our power of ten. We know it's -4, since we moved the decimal 4 places to the right.

$$3.789 \times 10^{-4}$$

Our answer is three and seven hundred, eighty-nine thousandths times ten to the negative four power.

HOT TIP

When you use scientific notation for large numbers, your exponent will always be **positive**. When you use scientific notation for numbers less than 1, your exponent will always be negative.

Quiz 2

1. What place is occupied by the numeral 5 in the decimal numbers below?

    a. 5,046.32

    b. 997.95

2. Perform the following additions and subtractions of decimal numbers.

    a. 123.5 + 234.6 + 98.09

    b. 9,780.65 − (234.5 − 198.2)

    c. 123.5 + (234.6 − 98.1)

3. Perform the following multiplications and divisions of decimal numbers.

    a. 54.2 × 19.1

    b. 61.8 ÷ 2.0 × 1.5

    c. 64.5 ÷ 0.5 × 9 ÷ 4.5

4. Round off these decimal numbers to the nearest hundredth.

    a. 35.045666667

    b. 9.9292929292

5. Write these decimal numbers in scientific notation.

    a. 0.00000789

    b. 12.0005

Fractions

1:04:17

We've looked at integers, decimals, and other animals. Now on to fractions. When you have a decimal, all the digits to the right of the decimal point represent a part of a whole number. This part of a whole number can also be expressed using fractions. But before we do some examples with fractions, we have to go over some terms.

1:04:43

Section A: Fraction Terms

A **fraction** is made up of two parts:

1:04:52

1:05:54

* ✖ The top number of a fraction is the **numerator**
* ✖ The bottom number is the **denominator**

$$\frac{3}{7}$$

In the fraction $\frac{3}{7}$, the 3 is the numerator and the 7 is the denominator. The denominator represents the whole, while the numerator represents the part of the whole that's present. One good way to explain fractions is to look at a pizza. Let's say there are 6 slices in 1 medium-sized pizza.

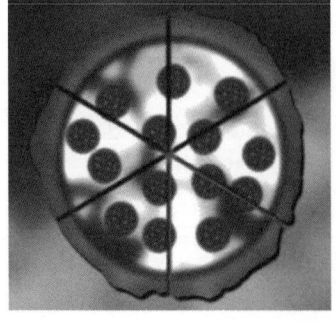

If all 6 slices are there, then the part and whole are the same. That means the value of the fraction is simply 1, because the numerator and the denominator are the same. You can divide the numerator, 6, by the denominator, 6, to get 1.

$$6 \div 6 = 1$$

$$\frac{6}{6} = 1$$

Fractions can have larger numerators than denominators too. If you go out and buy 9 slices, then you've got more than 1 whole pizza, since a medium-sized pizza only has 6 slices in it.

$$\frac{9}{6}$$

When this happens, the value of the fraction is greater than 1 because the part present is greater than the whole. You've got more slices than there are in a single medium-sized pizza.

Listen closely to this one: Sometimes fractions can be made up of different numbers, but actually have identical values.

All of the following fractions are actually equal to one another:

$$\frac{2}{6} = \frac{3}{9} = \frac{4}{12}$$

Fractions may seem kinda wacky, but hang on and this will all make sense. If we look at these fractions using a pie chart, then we can see that they all represent the same fraction of the pie.

Each of these fractions can be reduced to $\frac{1}{3}$.

How do you know you can reduce a fraction? Well, you can reduce a fraction when the numerator and the denominator have common factors. So what's a factor, you ask?

1:07:48

This is a vital term to have tucked away in your memory bank:

REMEMBER, a factor of a number is any number that can be multiplied by another number to get the original number.

It may sound confusing, but think of it this way. Take the number 6. There are 4 different whole numbers we can multiply by other whole numbers to get 6.

$$2 \times 3 = 6$$

$$1 \times 6 = 6$$

In other words, 6 has four factors: 1, 2, 3, and 6.

If you want to find the factors of a particular number, you just want 2 numbers you can multiply together to get that particular number. Now, let's return to our fractions and try to reduce them. We'll start by reducing $\frac{2}{6}$. Notice that both the numerator and the denominator of this fraction have 2 as a common factor. So we can reduce this fraction by dividing the numerator and the denominator by 2.

$$\frac{2 \div 2}{6 \div 2} = \frac{1}{3}$$

The resulting fraction is one-third. The value of the fraction has not changed--it has just been reduced. The value remains the same because both the numerator and denominator have been reduced by the same number, 2.

The same is true for $\frac{3}{9}$ and $\frac{4}{12}$. We can divide the first fraction by dividing the numerator and the denominator by 3, since both the numerator and denominator have 3 as a factor.

$$\frac{3 \div 3}{9 \div 3} = \frac{1}{3}$$

> **REMEMBER**, you reduce a fraction by finding a factor common to both the numerator and the denominator. Then you just divide the top and bottom of the fraction by the common factor.

The numerator and the denominator of the last fraction, $\frac{4}{12}$, can be reduced by dividing both parts by 4.

$$\frac{4 \div 4}{12 \div 4} = \frac{1}{3}$$

Put together your own dog by connecting this page with page 88.

So $\frac{2}{6}$, $\frac{3}{9}$, and $\frac{4}{12}$ are all equal. Pretty freaky, these fractions, eh?

Now that we've talked about the basics of fractions, let's put these puppies together and take 'em apart.

Section B: Adding and Subtracting Fractions

1:09:58

Now we'll tackle adding and subtracting fractions. But anytime you add or subtract fractions, you first have to check the denominators. We've got to take one important step if the denominators, or the bottom numbers of the fractions, are different. We'll look at fractions with common denominators first.

Let's say Wired Coffee Guy has 1 pot of coffee. Each pot has 6 equal cups of coffee in it.

If Wired Coffee Guy drinks 2 cups, what fraction of the original 6 cups of coffee does he have left now? Before we can figure out what fraction is left, we must first figure out what fraction of the original pot of coffee Wired Guy drank. He started with 6 cups of coffee, and those 6 cups made up the whole, or 1 pot of coffee. The fraction for that is $\frac{6}{6}$. Or, 6 cups of coffee out of 6. The amount of cups present was the same as the whole before Wired Guy started downing the caffeine.

$$\frac{6}{6} = 1$$

Mambo!

–Richard
Nixon

Each cup equals 1 cup out of the whole amount of coffee. That is, each cup represents one-sixth of the total amount of coffee brewed.

We can represent the two cups Wired Guy drank with the fractions $\frac{1}{6}$ and $\frac{1}{6}$.

$$\frac{1}{6} + \frac{1}{6} = ?$$

Notice that both fractions have a 6 in the denominator. That makes things easy. When we have the same denominators, like we do here, we just add the numerators together. We'll get to problems with different denominators in a second, so stay tuned.

> To add fractions with common denominators, just add the numerators, and leave the denominator alone.

When we add these two fractions, we see that

$$\frac{1}{6} + \frac{1}{6} = \frac{2}{6}$$

which tells us that Wired Coffee Guy drank $\frac{2}{6}$ of his original pot of coffee.

We can reduce this fraction by dividing the numerator and denominator by 2.

$$\frac{2 \div 2}{6 \div 2} = \frac{1}{3}$$

One-third of the pot of coffee is gone down Wired Coffee Guy's gullet.

Now we need to find out how much he left over. To figure that out, we'll subtract Wired Guy's 2 cups from the original pot of coffee.

> Just like when we add fractions with common denominators, we subtract fractions with common denominators by subtracting the numerators, while leaving the denominator untouched.

$$\frac{6}{6} - \frac{2}{6} = \frac{4}{6}$$

We could reduce this fraction further by dividing the 4 and 6 by their common factor, 2.

$$\frac{4 \div 2}{6 \div 2} = \frac{2}{3}$$

After 2 cups have been consumed, two-thirds of the original 6 cups of coffee still remain in the pot.

Now, if you check the denominators in the fractions you've got to add or subtract and they're different, then you've got to change them so that they have common denominators. Before we get to that, we have some more important terms for you to chew on.

Let's go back to factors.

1:13:05

Looking at the factors of a number will determine if the number is prime or composite. A **prime number** is any number whose only factors are 1 and itself. 7 and 19 are examples of prime numbers. We can multiply 1 times 7 to get 7, and 1 times 19 to equal 19, but that's it.

$$1 \times 7 = 7$$

$$1 \times 19 = 19$$

There are lots of prime numbers: 3, 5, 11, 17, 23, 29, 31, 37, 43, and on into infinity.

The second type of number is a **composite number**. This is any number that has more factors than just 1 and itself. 6 is a composite number.

$$2 \times 3 = 6$$

$$1 \times 6 = 6$$

This means that 1, 2, 3, and 6 are all factors of 6, since we can multiply each of these integers times another integer to get 6.

REMEMBER...

A prime number's only factors are 1 and itself. A composite number does have factors other than 1 and itself.

`1:14:16`

Now that you've got factors under your belt, we'll look at fractions that don't have common denominators. To add or subtract fractions without common denominators, we must find the lowest common denominator for both fractions. We want to make our denominators as small as possible.

> If the denominators in your problem are prime numbers, then the lowest common denominator is simply the first denominator times the second.

Why is this? It's because both numbers are as low as they can get. You're looking for the lowest number that both these denominators are factors of. So if we take the fractions $\frac{2}{3}$ and $\frac{4}{5}$, for instance, the lowest common denominator of 3 and 5 (both prime numbers) is 15. Both 3 and 5 are factors of 15.

We've got one more thing to throw at you.

> The lowest common denominator is also the product of the two denominators when both denominators are composite numbers and do not have any common factors other than one.

For example, both 8 and 9 are composite numbers.

$$1 \times 8 = 8 \qquad 1 \times 9 = 9$$
$$2 \times 4 = 8 \qquad 3 \times 3 = 9$$

The denominator of the fraction $\frac{3}{8}$, which is 8, has factors of 1, 2, 4, 8. The denominator of the fraction $\frac{4}{9}$, which is 9, has factors of 1, 3, 9. Notice that 1 is the only common factor of 8 and 9. Since 8 and 9 are composite and do not have any common fac-

tors other than 1, we find the two fractions' lowest common denominator by multiplying them together.

$$8 \times 9 = 72$$

The lowest common denominator of $\frac{3}{8}$ and $\frac{4}{9}$, then, is 72.

Let's say you want to find the common denominator for $\frac{1}{4}$ and

> Sometimes both your denominators are composite numbers and have common factors other than one. When this is the case, you find the lowest common denominator by figuring out the smallest factor common to both denominators.

$\frac{1}{10}$. Our first step is to find multiples of the larger of the two denominators, in this case, 10. To find multiples of 10, we simply multiply another number by 10. We're trying to find a multiple of 10 that has 4 as a factor. Since we're looking for the lowest common denominator possible, we'll multiply 10 by small numbers first.

$$10 \times 1 = 10$$

4 is not a factor of 10. There's no way to multiply 4 times another whole number to get 10. So we go to the next multiple.

$$10 \times 2 = 20$$

Aha. This should ring a bell. We know that 4 is a factor of 20.

$$4 \times 5 = 20$$

In other words, both 4 and 10 are factors of 20.

Remember, you're always looking for the lowest number that each denominator is a factor of. Since 20 can be divided by both 4 and 10, 20 is the lowest common denominator for 4 and 5.

$$20 \div 4 = 5$$

$$20 \div 10 = 2$$

Notice what happens if we continue to find multiples of 10.

$$10 \times 1 = 10$$

$$10 \times 2 = 20$$

$$10 \times 3 = 30$$

$$10 \times 4 = 40$$

40 is also divisible by 4 and 10.

$$40 \div 4 = 10$$

$$40 \div 10 = 4$$

But you are looking for the *lowest* common denominator, so even though you could use 40, you want to pick 20 because it's a smaller number.

Now you know how to find lowest common denominators. But we can't just switch the denominators and go on. We have to deal with the numerators, too. **We take each fraction separately, and multiply its numerator by whatever number we chose to multiply its denominator by (when we figured out the lowest common denominator).** We'll make this clearer by adding a couple of fractions.

$$\frac{7}{12} + \frac{5}{18} = ?$$

The first step is to find the lowest common denominator. We can't add fractions unless the denominators are the same. Both 12 and 18 are composite numbers, and they have factors other than 1. So we look for the lowest common denominator by finding the smallest number that has both 12 and 18 as its factors. We start by finding multiples of the larger number, 18.

$$18 \times 1 = 18$$

12 isn't a factor of 18. We can't multiply 12 by any number to equal 18. So we keep finding multiples of 18.

$$18 \times 2 = 36$$

12 is a factor of 36 because we can multiply 12 by 3 to get 36. Now we can see that the smallest number that has both 12 and 18 as factors is 36.

$$12 \times 3 = 36$$

$$18 \times 2 = 36$$

Once we've found that the two fractions' common denominator is 36, we can multiply the numerator and the denominator of each of the fractions to make their denominators the same.

REMEMBER, as long as we multiply the numerator and denominator of a fraction by the same number, the value of the fraction doesn't change.

Oh, flesh, flesh, how art thou fishified.

—Romeo and Juliet

We can make the denominators of both fractions the same without altering the value of the fractions. Let's start with seven-twelfths. Since we know that 12 times 3 equals 36, we multiply both parts of the fraction by 3.

$$\frac{7 \times 3}{12 \times 3} = \frac{21}{36}$$

Now we'll transform the other fraction. The denominator is 18, and we know that 2 times 18 equals 36, so we'll multiply the numerator and denominator by 2.

$$\frac{5 \times 2}{18 \times 2} = \frac{10}{36}$$

Our fraction is now ten over thirty-six, which means that both of our denominators are the same. We haven't changed the value of either fraction, just altered them so the denominators are the same.

$$\frac{21}{36} + \frac{10}{36} = ?$$

Our fractions now have the same denominators, so we can add the fractions together. We add the numerators and leave the denominators alone.

$$\frac{21}{36} + \frac{10}{36} = \frac{31}{36}$$

simplicity

31 and 36 have no common factors, so we can't reduce the fraction any further. Our answer is in its simplest form.

We won't do one here, but we subtract fractions the same way we add them--that is, by first finding the lowest common denominator.

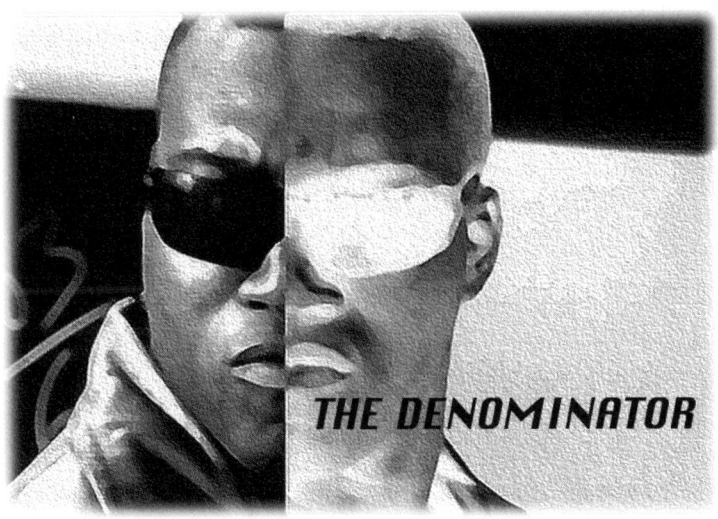

THE DENOMINATOR

REMEMBER,

WHEN YOU WANT TO FIND THE LOWEST COMMON DENOMINATOR FOR FRACTIONS WITH DIFFERENT DENOMINATORS, THERE ARE 3 POSSIBLE SCENARIOS:

(1) If both denominators are prime, just multiply them together.

(2) If your denominators are composite and have no common factors other than 1, you multiply the denominators together.

(3) If the denominators are both composite and have factors other than 1, you find the lowest possible number that both denominators are factors of. Then multiply the numerator and the nominator of each fraction so that your denominators are the same.

1:22:54

Section C: Multiplying and Dividing Fractions

Multiplying fractions is easier than adding or subtracting them because you don't need to worry about having a common denominator.

To multiply fractions, you multiply the numerators together to get your product's numerator, and then you multiply your denominators together to get your product's denominator.

We'll multiply four-thirds times five-eighths.

$$\frac{4}{3} \times \frac{5}{8} = ?$$

First we multiply the numerators together.

$$4 \times 5 = 20$$

Then we multiply our denominators together.

$$3 \times 8 = 24$$

Now we've got 20 as our numerator and 24 as our denominator.

$$\frac{20}{24}$$

We can reduce this fraction because both numbers in the fraction are divisible by 4. (Friendly Reminder: Always do the same thing to the numerator as you do to the denominator, to make sure that the value of the fraction stays the same.)

$$\frac{20 \div 4}{24 \div 4} = \frac{5}{6}$$

A 9 is a 6 with self-confidence.

– Fortune cookie

Our simplified product looks like this.

$$\frac{4}{3} \times \frac{5}{8} = \frac{20}{24} = \frac{5}{6}$$

Now we'll look at dividing fractions.

Remember we said that division is the reverse of multiplication. This is a neat little trick that makes dividing fractions a whole lot easier. If you can multiply fractions, you're all set to divide them, too.

$$\frac{8}{3} \div \frac{4}{6} = ?$$

Instead of trying to figure out how many times $\frac{4}{6}$ goes into $\frac{8}{3}$, we use the fact that division is the reverse of multiplication to find the answer.

Watch closely: instead of dividing by four-sixths, we flip it to make it six-fourths.

$$\frac{4}{6} \rightarrow \frac{6}{4}$$

$\frac{6}{4}$ is known as the reciprocal of $\frac{4}{6}$. The **reciprocal** of a fraction is simply the fraction flipped over. For example, the reciprocal of eight-tenths is ten-eighths.

$$\frac{8}{10} \rightarrow \frac{10}{8}$$

Let's go back to our division problem. We flipped the fraction $\frac{4}{6}$ to make it $\frac{6}{4}$. Our next step is to multiply the two fractions.

$$\frac{8}{3} \times \frac{6}{4} = ?$$

We multiply $\frac{8}{3}$ times $\frac{6}{4}$ in the same way we multiplied fractions earlier. We multiply the numerators together to find the product's numerator, then we multiply our denominators together to get the product's denominator.

$$\frac{8 \times 6 = 48}{3 \times 4 = 12}$$

Our product is forty-eight twelfths.

$$\frac{48}{12}$$

We need to take one more step to finish off this problem. Forty-eight-twelfths can be reduced because both numbers are divisible by 12. So, we divide both the numerator and denominator by 12.

$$\frac{48 \div 12}{12 \div 12} = \frac{4}{1}$$

Our answer is $\frac{4}{1}$, or 4.

Why just 4? Because dividing the numerator, 4, by the denominator, 1, is 4. So the answer to $\frac{8}{3}$ divided by $\frac{4}{6}$ is 4.

$$\frac{8}{3} \div \frac{4}{6} = 4$$

REMEMBER...

(1) To multiply fractions you multiply the numerators together then the denominators together. Then you reduce the fraction if possible.

(2) To divide fractions, first you have to flip over the fraction you're dividing by to find its reciprocal. Then you just multiply the first fraction by the flipped-over fraction and simplify your answer.

STUDY SIDEKICK

`1:.26:47`

Section D: Mixed Numbers

The last thing we'll discuss in this section on fractions is a mixed number. A **mixed number** is any number that contains an integer and a fraction, like three and one-sixth.

$$3\frac{1}{6}$$

`1:27:26`

Anytime you're working with mixed numbers, you first want to convert the mixed number into an improper fraction. What's an improper fraction? Well, an **improper fraction** is one which has a larger numerator than denominator. It's kinda top-heavy. We'll explain what we mean by adding a couple of these mixed numbers.

$$3\frac{1}{6} + 2\frac{1}{3} = ?$$

VIDEO NOTES

Our first step in this problem is to rewrite $3\frac{1}{6}$ and $2\frac{1}{3}$ to make them improper fractions. There are two ways to convert mixed numbers into improper fractions. We'll convert $3\frac{1}{6}$ the first way.

The first thing you need to do is convert the whole number, 3, into an improper fraction. This improper fraction should have the same denominator as the fraction-part of your original number, $3\frac{1}{6}$. Since 6 is the denominator, you want to make the whole number, 3, a fraction that has 6 as its denominator. How do you do this? The best way is to multiply your whole number times the denominator in your fraction.

$$3 \times 6 = 18$$

18 is your numerator and 6 remains the denominator. You haven't changed the value of your fraction. 3 is the same thing as $\frac{18}{6}$ because 6 can be divided into 18 three times.

$$\frac{18}{6} = 3$$

Now you just add $\frac{18}{6}$, your whole number, to $\frac{1}{6}$, your fraction.

$$\frac{18}{6} + \frac{1}{6} = \frac{19}{6}$$

Nineteen-sixths has the same value as three and one-sixth.

$$3\frac{1}{6} = \frac{19}{6}$$

STUDY SIDEKICK

Our mixed number is now condensed into a nice little package: an improper fraction.

Now we'll convert $2\frac{1}{3}$ into an improper fraction, using another method. We find the numerator of the improper fraction by multiplying the whole number, 2, times the fraction's denominator, 3, and then add the numerator, 1.

First we'll multiply our whole number times the denominator.

$$2 \times 3 = 6$$

Then we just add the numerator, 1.

$$6 + 1 = 7$$

The 7 is the numerator in our improper fraction. The denominator stays the same.

$$2\frac{1}{3} = \frac{7}{3}$$

Either way you convert the whole number, you get the same improper fraction.

Now we can substitute improper fractions for our mixed numbers. The updated problem looks like this.

$$\frac{19}{6} + \frac{7}{3} = ?$$

We add the fractions just like we have in the past. We first find the lowest common denominator of 6 and 3. To do so, we start by determining the factors of 6.

$$1 \times 6 = 6$$

$$2 \times 3 = 6$$

The factors of 6 are 1, 2, 3, and 6. Since 3 is the other denominator, and it's a factor of 6, we know that 6 is the denominator. So $\frac{19}{6}$ stays the way it is, since it already has 6 as a denominator. To give $\frac{7}{3}$ a denominator of 6, we multiply both the numerator and denominator by 2.

$$\frac{7 \times 2}{3 \times 2} = \frac{14}{6}$$

That gives us a new fraction of fourteen-sixths. Now that the denominators are the same, we can add the two fractions.

$$\frac{14}{6} + \frac{19}{6} = \frac{33}{6}$$

We're almost done. We can reduce this fraction because both numbers in the fraction are divisible by 3.

$$\frac{33 \div 3}{6 \div 3} = \frac{11}{2}$$

We can either leave the fraction this way or convert it back to a mixed number. We convert it back to a mixed number by dividing the denominator into the numerator.

Gosh darn
these tiny
bugs!

– Jennie
Halfant

$$2\overline{)11} \atop \begin{array}{r} 5 \\ \underline{10} \\ 1 \end{array}$$

The remainder becomes the numerator of the new fraction, and 5 becomes the whole-number part of the mixed number.

$$3\frac{1}{6} + 2\frac{1}{3} = 5\frac{1}{2}$$

That's it for fractions!

Quiz 3

1. What are the numerators and the denominators of these fractions?

   a. $\dfrac{12}{13}$

   b. $\dfrac{5}{7}$

2. Add or subtract the following fractions.

   a. $\dfrac{1}{2} + \dfrac{3}{5}$

   b. $\dfrac{8}{9} - \dfrac{1}{3} + \dfrac{5}{6}$

3. Multiply or divide these fractions.

   a. $\dfrac{1}{2} \times \dfrac{3}{4}$

   b. $\dfrac{5}{7} \div \dfrac{25}{70}$

4. Perform the indicated operations on the following mixed numbers.

   a. $1\dfrac{1}{3} - \dfrac{5}{6}$

   b. $1\dfrac{3}{7} \times 5\dfrac{3}{4}$

Ratios and Percents

For the rest of the workbook, we're going to take all this stuff we've been learning and apply it to some real-life examples.

Section A: Ratios

Ratios are the first application we're going to look at. A **ratio** is a relationship between quantities or amounts. This relationship can be expressed in a number of ways. Like fractions, ratios tell you how much you have in comparison to the whole.

We'll go back and take cups of coffee as an example. This time, there's 11 cups on the counter. 5 are regular-strength coffee, 4 are espressos, and 2 are cappuccinos. We'll put these into a ratio. You indicate that numbers are in a ratio by placing colons between the amounts present, like the number of cups of coffee, espresso, and cappuccino.

The ratio is 5 quantities of regular coffee, 4 quantities of espresso, and 2 quantities of cappuccino. We stick colons in between each of these quantities (or parts of the whole that are present).

$$5 : 4 : 2$$

This means that the ratio of regular coffee to espresso to cappuccino is 5 to 4 to 2. Notice that there are a total of 11 cups.

$$5 + 4 + 2 = 11$$

The 11 represents the whole in this case.

If we looked at this ratio in terms of fractions, then 11 would be the denominator, since it represents the whole. The numerator would be either 5 coffees or 4 espressos or 2 cappuccinos. In other words, the regular coffee represents $\frac{5}{11}$ of the total cups there. $\frac{4}{11}$ of the total is represented by the espresso, and $\frac{2}{11}$ of the whole is cappuccino.

$$\frac{5}{11} = \text{coffee} \quad \frac{4}{11} = \text{espresso} \quad \frac{2}{11} = \text{cappuccino}$$

We'll do one more example of ratios. In a recent poll, 27 students liked the idea of having two jujitsu masters engage in combat while their professor lectured, while 28 students preferred the less violent game of handball. 55 students were surveyed in all.

27 students + 28 students = 55 students

So, out of a class of 55 students, 27 prefer jujitsu masters as classroom entertainment.

We can put this ratio into fraction form. The denominator represents the whole (the 55 students sampled). The numerator is the part present, or the 27 students who chose jujitsu masters.

$$\frac{27}{55}$$

We can do the same thing for those 28 students who prefer handball.

$$\frac{28}{55}$$

This fraction represents the 28 out of 55 students who prefer handball. And that's all there is to it!

Section B: Percents

Percents are the final application we will discuss. You've probably seen percentages in many places before. Percent comes from Latin and means "per hundred." Basically, it's just a number over 100. For example, 50% is equal to $\frac{50}{100}$. We can also reduce this to the fraction $\frac{1}{2}$, or the decimal .5. They all represent the same amount.

$$50\% = \frac{50}{100} = \frac{1}{2} = 0.5$$

Let's say that on your test you're asked to convert a decimal, such as .35, into a percent. What do you do? You simply move the decimal point two places to the right, and insert a percent symbol.

$$0.35 \rightarrow 35\%$$

Why, you ask? Because all you're doing is multiplying 0.35 times one hundred. Remember, each zero represents a digit place. So if we move the decimal point two places, it's like we're multiplying the number by 100. Here's another example. We'll convert this number into a percent.

$$1.695$$

We move the decimal place 2 digits to the right.

$$1.695 \rightarrow 169.5\%$$

111

STUDY SIDEKICK

The decimal number has been trans-
formed into a percent. And when you're
doing percent problems, write the sign
like this to brighten someone's day.

REMEMBER...

To convert a fraction into a percent, you
divide the denominator into the numerator. Once you have a decimal, move the
decimal 2 places to the right and insert
the percent symbol.

Let's go back to our classroom example. Remember, a poll was
taken recently showing that 27 people out of 55 prefer jujitsu
masters in their classroom. We want to find out what percent-
age that makes. The first step is to convert the fraction $\frac{27}{55}$
into a decimal. We know that we can convert the fraction $\frac{27}{55}$
to a decimal by dividing 55 into 27.

$$27 \div 55 = 0.490909$$

Okay, we've got our decimal. To transform this number into a
percent, we move the decimal point 2 places to the right and
insert the percent symbol.

$$.490909 \to 49.0909\%$$

We can round 49.0909% off to 49%. We can now say that 49% of
the students prefer jujitsu masters over handball.

VIDEO NOTES

This ends our study of percents and the basics of numbers and mathematical operations. But don't cry! Perk up, little buckaroo. There's lots more math to come! We got more quizzes! Tests! Other Important Stuff! A Recipe!! A Crossword Puzzle!! And Lots More!!

MR. GUMBY:
My brain
hurts.

BRAIN
SPECIALIST:
Well, we're
just going to
have to take
that thing out,
Mr. Gumby.

*– Monty
Python's
Flying Circus*

Quiz 4

1. Find the following ratios.

    a. Frank looks in his freezer and sees that he has 100 fudgecicles, 40 grape popsicles, and 60 cases of motor oil. As you can tell, Frank's got a big freezer. What is the ratio of the objects in Frank's freezer? (Reduce to lowest possible ratio.)

    b. Marsha supervises a fleet of cars that her company rents. If they have 6 stretch Cadillacs and 4 Corvettes made by General Motors, and 36 Lincoln Town Cars and 24 Broncos made by Ford Motor Company, what is the ratio in the fleet of vehicles made by GM to vehicles made by Ford?

2. Convert the following numbers into percents:

    a. $\frac{1}{4}$

    b. 2

A Joltin' Tale of Java
to Jangle Your Nerve and Jitter Your Verve

I once had a boss named Sally
who drank more coffee than Einstein could tally.
She swigged and she swallowed,
she slurped and she burped.
She gulped it with glorious gusto.

Sally zoomed from building to building,
always wielding her mug.
Nothing could faze her, could flip her, could freak her.
Not even burnt coffee made her bug.
When her coffee would steam,
she'd just dump in more cream,
and then oh, how you'd see Sally beam.

Sally ricocheted off ceilings;
she bounced from walls and wallpaper peelings.
She tore through corridors
with a mighty roar: "Coffee! Coffee! Give me more!"

One fine but fateful day,
Sally summoned me in a feverish way.
She was shaking and quaking
because I drank all the coffee she was making.
I was about to express my remiss
when my coffee-breath escaped with a hiss.
She shivered with glee and said, "Come closer to me."
I was near her ear when I felt a flush of fear:
"Oh my dear fellow," I heard her bellow.
Frantic, I tried to flee.
But before I knew it she'd swallow me!
As I slipped and sledded down the
caffeine-coated path to my doom,
I heard her sigh and whisper, "Oh my..."
So now here I sit in her stomach pit,
listening to her mighty roar:
"Coffee! Coffee! Give me more!"

TRIPE, LOUISIANA STYLE

A hearty sauté that blends good 'ol cajun flavor with an often overlooked animal organ.

The Ingredients:

- 2 lb. tripe
- sugar
- *Tony's Seasonings* to taste
- can of diced tomatoes
- 2 onions
- green pepper
- 1 cup celery
- 2 garlic cloves
- 1 cup diced, cooked ham

The Cooking Process:

1 - Wash tripe thoroughly. Cut into strips and place into pan filled with cold water. Bring water to a boil and add salt and sugar. Cover and simmer for 3 hours, or until tripe is tender. Drain.

2 - Add tomatoes, green pepper, onion, celery, garlic, and *Tony's* to tripe. Cover and simmer for 20 minutes.

3 - Add ham. Cover and simmer for 15 more minutes. Serve hot over rice or by itself. Serves 4-6 people.

PRACTICE EXAM

1. Name the position of the 7 in each of the numerals below:

    a. 0.736

    b. 7,562.65

    c. 137

    d. 63.467

    e. 7,530,000

    f. 0.00007

2. Add the following integers:

    a. 729 + 1,036

    b. 4,598 + 54 + 366

    c. 34 + 256,999 + 56 + 4 + 1,101

    d. 123 + 321 + 4,444

    e. 46,789 + 56,301 + 67,943 + 11,009

3. Subtract the following integers:

    a. 123 − 45

    b. 45,678 − 34,908

    c. 350,000 − 47,987

    d. 358,000 − 47,987

    e. 9,876 − 25

4. Multiply the following integers:

     a. 56×78

     b. $43,675 \times 6$

     c. 469×372

     d. 144×233

     e. $(21 \times 34) \times 55$

5. Divide the following integers:

     a. $1,625 \div 25$

     b. $15,411 \div 33$

     c. $74,539 \div 131$

     d. $5,334 \div 9$

     e. $1900 \div 4$

6. Evaluate the following:

     a. 2^5

     b. 1^9

     c. 13^2

     d. 10^3

     e. 7^0

7. Simplify the following expressions:

    a. $(5 + 7) - 3 + 12 \div 2$

    b. $(5 + 7 - 3 + 12) \div 12$

    c. $82 + 2[41 - 3(6 - 4)]$

    d. $56{,}274 - [12{,}000 + (3)(4{,}321) + 32675 \div 5]$

    e. $56 \div 7 + (33 + 2)(14 - 9) \div 5(32 - 23)$

(Solving this final equation will be easier if you put it into the form of a fraction)

8. A new arena has 25,679 seats for which it sells tickets at $15.00 and a total of 30,000 seats for which it may sell tickets.

    a. If the arena sells all the $15.00 tickets for a performance, how much will it gross for that performance?

    b. If the arena sells all the $15.00 tickets, and then sells half the remaining tickets for $25.00 per ticket, how much will it gross for that performance?

    c. If the arena sells 35% of the $25.00 tickets and 85% of the $15.00 tickets for a certain performance, how much will it gross for that performance?

9. A student is ordering pizzas and sodas for his study group. The pizzas are $6.00 each (six slices in a pizza) or $1.50 per slice. The sodas cost $1.25 each. There are four men and five women in the study group.

    a. If each man eats 4 slices of pizza, each woman eats 2 slices of pizza, and each student drinks 2 sodas, how much will the bill come to (before tip)?

b. If the tip is 15% of the bill, how much is the tip?

c. If the students split the total bill (bill plus tip) equally, how much should each person chip in?

10. A student is checking her car for gas mileage performance during a recent three-day trip. She notes her beginning and ending odometer readings as well as the gas required each time she stops to fill up the gas tank.

Starting Odometer	Ending Odometer	Gallons of Gas Used
12,500.9	12,832.4	13.9
12,832.4	13,057.6	9.5
13,057.6	13,345.6	12.3

a. Gas mileage is found by dividing the number of miles driven by the number of gallons of gas used during the drive. Find the student's gas mileage for the three different occasions when she stopped for gas (shown in the table above).

b. If she purchased gas at an average of $1.46 per gallon, what was her cost for the gas used during the trip?

11. Write the following numbers in scientific notation:

a. 647

b. 2,345,543,247,000

c. 0.00000563

d. 8,011,432.007

e. 100

12. Reduce the following fractions to their lowest terms:

a. $\dfrac{25}{75}$

b. $\dfrac{130}{26}$

c. $\dfrac{17}{51}$

d. $\dfrac{210}{9240}$

e. $\dfrac{105}{1155}$

13 Solve the following fractions, making sure that all answers have been reduced to their lowest terms:

a. $\dfrac{1}{4} + \dfrac{5}{7}$

b. $\dfrac{1}{3} - \dfrac{12}{51} + \dfrac{15}{18}$

c. $4 + \dfrac{2}{5} + \dfrac{1}{12}$

d. $\dfrac{3}{4} - \dfrac{21}{32} + \dfrac{1}{17}$

> Your breath
> stinks with
> toasted
> cheese.
>
> – *Henry IV,*
> *Part 2*

14. Multiply the following:

a. $\dfrac{2}{5} \times \dfrac{3}{7}$

b. $\dfrac{3}{7} \times \dfrac{6}{4} \times \dfrac{1}{11}$

c. $7\dfrac{2}{3} \times \dfrac{5}{8}$

d. $\dfrac{1}{4} \times \dfrac{6}{73} \times \dfrac{73}{18}$

15. Divide the following:

a. $\dfrac{2}{3} \div \dfrac{1}{5}$

b. $\dfrac{5}{8} \div \dfrac{8}{31}$

c. $\dfrac{11}{4} \div \dfrac{3}{4}$

d. $\dfrac{16}{25} \div \dfrac{8}{9}$

e. $\left(2 + \dfrac{3}{8}\right) \div \left(1 + \dfrac{4}{7}\right)$

16. Find the decimal and percent forms of the following numbers:

a. $\dfrac{5}{8}$

b. $\dfrac{2}{9}$

c. $1\dfrac{2}{7}$

d. 42

e. $\dfrac{11}{6}$

17. If a student has 40 pages of notes from his math class and 15 pages of notes from his history class, what is the ratio of math notes to history notes?

18. What is 62.5% of 1,142, if you round the answer to the units place?

19. Deviant University has a freshman class of 4,580 students. If 30% of the students want to take freshman English at 8 a.m., and there are 5 classes taught at that hour with no more than 100 students in each class, how many students who want to get into that early English class will not be able to?

20. Ms. Dina Soar, a paleontologist, has uncovered a total of 840 fossilized bones in her latest dig. If she has identified that

10% are from velociraptors and $\frac{1}{4}$ are from pteranodons, how many does she have left to identify? What is the ratio of the identified bones to the unidentified bones?

21. Dr. James Pearson, a famous astronomer, has discovered a new heavenly body that is approximately 0.25 light-year away. If a light-second is equal to 186,000 miles, and there are 60 light-seconds in a light-minute, and 60 light-minutes in a light-hour, and 24 light-hours in a light-day and 365.6 light-days in a light-year, how many miles away is Dr. Pearson's heavenly body? (Use scientific notation and round your answer to three decimal places.)

22. Two years ago there was 1 coffee bar for every 4 blocks in Washington, D.C. Now, there are 3 coffee bars on every block of that same city. What is the ratio of the coffee bars today to the coffee bars 2 years ago?

23. A student working his way through college is paid at an hourly rate of $7.62. After he gains some experience, he receives a raise of 5%. Two months later, he, like all the other employees, received a cost-of-living raise of 3.1%. What was his hourly rate after both raises?

OTHER IMPORTANT STUFF

Stuff 1: Properties of Addition and Multiplication

The rules (or properties) of addition and multiplication are true for all real numbers of any kind--integers, fractions, decimals, mixed numbers, whatever. These properties are the basis for basic math, so it is good to learn them (or at least have them around as a reference for whenever you're not sure whether an operation you want to do is valid). Their names may sound formal, but the laws themselves are pretty simple.

Commutative Property of Addition: The sum of a pair of numbers does not change if you change the order in which you add the numbers.

$$\text{ex. } 1 + 2 = 2 + 1 = 3$$

Associative Property of Addition: The sum of a group of numbers that are to be added together does not change if you add them together as different pairs.

$$\text{ex. } 1 + (2 + 3) = 1 + 5 = 6 \text{ and } (1 + 2) + 3 = 3 + 3 = 6$$

STUDY SIDEKICK

Additive Identity: This is a number that, when added to any other number, does not change the other number. For addition of real numbers, the additive identity is 0.

> *You are as rheumatic as two dry toasts.*
>
> *–Henry IV, Part 2*

ex. $2 + 0 = 2$

Additive Inverse: A number that, when added to its inverse, results in zero. For real numbers, you find the additive inverse for any number by multiplying that number by -1.

ex. $5 + (-5) = 5 - 5 = 0$

Commutative Property of Multiplication: The product of a multiplication problem is not changed if you change the order of its factors.

ex. $2 \times 3 = 3 \times 2 = 6$

Associative Property of Multiplication: The product of 3 or more factors is not changed if you alter the grouping of the factors to carry out the operation.

ex. $2 \times (3 \times 4) = 2 \times 12 = 24$ and $(2 \times 3) \times 4 = 6 \times 4 = 24$

Multiplicative Identity: This is a number that, when multiplied by any other number, results in the other number. For real numbers, the multiplicative identity is 1.

ex. $2{,}435 \times 1 = 2{,}435$

Multiplicative Inverse: A number that, when used to multiply another number (except 0), results in a product of 1. For real numbers, the multiplicative inverse of a number is its reciprocal (a fraction with 1 in the numerator and the number in the denominator).

$$\text{ex. } 2 \times \frac{1}{2} = \frac{2}{1} \times \frac{1}{2} = \frac{2}{2} = 1$$

Multiplicative Property of Zero: Any number multiplied by zero is zero.

$$\text{ex. } 1.8 \times 1047 \times 0 = 0$$

Distributive Property: This one can get a little confusing, but it's very very important in algebra. It allows you to combine the operations of addition and multiplication, like this:

$$2 \times (3 + 4) = (2 \times 3) + (2 \times 4) = 6 + 8 = 14,$$

(also notice that $2 \times (3 + 4) = 2 \times 7 = 14$),

$$\text{or } (2 + 3) \times 4 = (2 \times 4) + (3 \times 4) = 8 + 12 = 20$$

(also notice that $(2 + 3) \times 4 = 5 \times 4 = 20$)

Stuff 2: Laws of Exponents

These laws hold true for all real numbers:

SYMBOLS	TRANSLATION
$4^2 \times 4^3 = 4^{2+3} = 4^5$	When the bases (here, the 4's) are EXACTLY ALIKE, multiply them by adding their exponents.
$(4^2)^3 = 4^{2 \times 3} = 4^6$	When you are raising an exponent to another power, multiply the exponents.
$(4 \times 5)^2 = 4^2 \times 5^2$	When you are raising a product to a power, raise **EACH FACTOR** of the product to that power.
$\left(\dfrac{4}{5}\right)^2 = \dfrac{4^2}{5^2}$	When you are raising a fraction to a power, raise **BOTH** the numerator and the denominator to that power.
$4^0 = 1$	Any number (or expression) that is not zero, when raised to the zero power, is equal to 1.
$4^{-2} = \dfrac{1}{4^2}$	Any number (or expression) that is not equal to zero, if raised to a negative power, is the reciprocal of that number (or expression) raised to the positive power.
$4^3 \div 4^2 = 4^{3-2} = 4^1 = 4$	When the bases (here, the 4's) are **EXACTLY ALIKE** (and not equal to zero), divide by subtracting their exponents.

Stuff 3: Divisibility Rules

ANY WHOLE NUMBER THAT:	IS DIVISIBLE BY:
Ends in 0 or 2 or 4 or 6 or 8	2
Has the sum of its digits divisible by 3	3
Ends in 0 or 5	5
Is divisible by 2 AND by 3	6
Has the sum of its digits divisible by 9	9
Ends in 0	10

These rules are tricks for figuring out even divisors (also known as factors) of some commonly used numbers. Memorizing them will really help you fly through your homework, quizzes and exams.

The following numbers are some of the many prime numbers that have NO factors except for themselves and 1:

1, 2, 3, 5, 7, 11, 13, 17, 19, 23, 29, 31, 37, 41, 43, 47, 53, 59

Division by zero is undefined, so as far as math is concerned, you can't do it. So don't do it. It is always wrong. If you try to do it, you will receive a guaranteed zero points for trying it on any test by any teacher anywhere. If you continue to divide by zero you will be forced to listen to Jefferson Starship and drink jars of pickle juice until you repent.

Stuff 4: How To Deal with Signed Numbers

OPERATION	POLARITY OF SIGN	WHAT TO DO
ADDITION		
	Both Positive	Add the unsigned numbers, keep the + sign: $5 + 7 = 12$
	Both Negative	Add the unsigned numbers, keep the $-$ sign: $-5 + (-7) = -(5 + 7) = -12$
	Positive and Negative	Subtract the smaller unsigned number from the larger unsigned number. Use the sign that belonged to the larger unsigned number in the result: $5 + (-7) = -(7 - 5) = -2$
SUBTRACTION		
	Both Positive	Subtract one number from another by adding the inverse of the number to be taken away: $2 - 5 = 2 + (-5)$
	Both Negative	Add the inverse of the number to be taken away. So, $-2 - (-5) = -2 + 5 = +(5 - 2) = +3$ and $-5 - (-2) = -5 + 2 = -(5 - 2) = -3.$

130

| Positive and Negative | Treat these just like the others: add the inverse of the number to be taken away. |

$$-5 - (-2) = -5 + 2 =$$
$$-(5 - 2) = -3 \text{ and} -2 - (-5)$$
$$= -2 + 5 = +(5 - 2) = +3$$

MULTIPLICATION

| Both Positive | Just multiply the numbers. Positive times positive is always positive. |

| Both Negative | Multiply the unsigned numbers and make the product's sign positive. Negative (-) times negative (-) is always positive (+). But be careful: negative times negative times negative is negative. Why? Because (-) times (-) equals (+) but you have to multiply this (+) times one more (-), which leaves you with a (-)! So stay awake. |

| Positive and Negative | Multiply the unsigned numbers and make the product's sign negative. |

$$(-2) \times 3 -$$
$$-(2 \times 3) = -6.$$

DIVISION

Both Positive	Just divide the numbers. Positive divided by positive is always positive.
Both Negative	Divide the unsigned numbers and make the quotient's sign positive. Negative divided by negative is always positive.
Positive and Negative	Divide the unsigned numbers and make the product's sign negative. So, $(-2) \div 3 = -(2 \div 3) = -2 \div 3$. And, $3 \div (-2) = -(3 \div 2) = -1\frac{1}{2}$.

EXPONENTIATION

Both Positive	Raise the base to the power just as you do with unsigned numbers. $5^2 = 5 \times 5 = 25$ and $2^3 = 2 \times 2 \times 2 = 8$
Both Negative	Raise the base to the power and find its reciprocal. Then, make the answer's sign negative if the power is an odd number and positive if the power is an even number. $(-3)^{-2} = \frac{1}{(-3)^2} = \frac{1}{(3 \times 3)} = \frac{1}{9}$ and $-3^{-3} = -\frac{1}{(3 \times 3)} = -\frac{1}{27}$

Base Is Negative, but Power Is Positive	Raise the unsigned base to the power and make the answer's sign positive. $(-2)^2 = 2^2 = 2 \times 2 = 4$ (This is because multiplication of two negatives results in a positive.)
Base Is Positive, but Power is Negative	Raise the base to the power and find its reciprocal. Then make the answer's sign positive $3^{-2} = \dfrac{1}{3^2} = \dfrac{1}{(3 \times 3)} = \dfrac{1}{9}$

Whenever you're in doubt about whether your answer is positive or negative, DRAW A NUMBER LINE and walk your fingers back and forth counting off the numbers until you come to your answer. It will help clear up some of these pesky operations with signed integers, especially those that use subtraction.

Stuff 5: How To Solve Word Problems

1. Read the problem. Think about the quantities it gives you and the ones you are asked to find. Make sure the units are the same for both.

2. Sketch out the problem if it is complicated, or about areas or perimeters, to help you visualize what you are being asked to find.

3. Translate the words into mathematical operations (and, sometimes, equations).

4. Do the operations according to the rules you've learned, and if you have an equation, solve it according to the rules you've learned about working with equations.

5. Check the solution for arithmetic errors.

6. Check the solution against the word problem to be sure it is reasonable and that your translation was correct. For example, if your answer is in miles and the problem asked you to find an area, which is always in square units, something is wrong. It's even worse if you are trying to find the number of tickets sold and you end up with an answer in miles. In this case, better check the time and see just how long you've gone without sleep.

Here is a list of keywords that you can use to help you translate a word problem:

ADDITION	SUBTRACTION	MULTIPLICATION	DIVISION
added to	decreased by	area	divided (by
total	take away	times	or into)
and	deduct	multiplied by	each
increased by	difference	of	per
more than	less than	product of	quotient
perimeter	minus		remainder
plus	subtracted from		
sum			

(Here is an example:

Near the end of a long and successful season, the members of a soccer team decide to buy a lot of liquid butter and dump it on their coach after their last game. One teammate has a 10-gallon jug they can use, so the team decides to fill it completely. They can walk over to the nearest convenience store and buy individual 20-oz bottles for $0.85, or they can drive a couple of miles to a grocery store and get it in liter bottles for $0.65 each. Assuming they use the coach's car, so no one on the team is out for gas, answer the following:

a. How much will the liquid butter cost if the team buys all of it at the convenience store?

b. How much will the liquid butter cost if the soccer team buys all of it at the grocery store?

c. How much will the liquid butter cost if they split up and buy $\frac{1}{3}$ of it at the convenience store and $\frac{2}{3}$ of it at the grocery store?

The Solution:

First, we read the problem and see that we have to buy enough liquid butter to fill a 10-gallon jug. This means they may have to buy a little more than 10 gallons of liquid butter, since it doesn't come in gallon sizes at either store. But they can't buy less than 10 gallons because then they couldn't fill up the jug.

Then, we notice that we have three kinds of volume units in this problem: gallons, ounces, and liters. So we will have to convert the 10 gallons to an equivalent number of ounces and liters. That way, we can find out how many bottles of each the team needs.

First, we'll find out how many ounces are in 10 gallons. There are 4 quarts per gallon, and 2 pints per quart. That means that there are 8 pints per gallon, because we multiply 2 (the number of pints in a quart) times 4 (the number of quarts in a gallon).

$$
\begin{array}{rl}
4 & \text{quarts in a gallon} \\
\times\,\underline{2} & \underline{\text{pints per quart}} \\
8 & \text{pints in a gallon}
\end{array}
$$

OTHER STUFF

We also know that there are 16 ounces per pint. So we multiply 8, the number of pints per gallon, times 16, the number of ounces per pint.

$$
\begin{array}{r}
8 \quad \text{pints per gallon} \\
\times\ 16 \quad \underline{\text{ounces per pint}} \\
128 \quad \text{ounces per gallon}
\end{array}
$$

Now we know that there are 128 ounces in a gallon. We're trying to find the number of ounces in 10 gallons. Our last step is to multiply the 128 ounces in each gallon times 10 gallons.

$$
\begin{array}{r}
128 \quad \text{ounces in a gallon} \\
\times\ 10 \quad \underline{\text{gallons}} \\
1280 \quad \text{ounces in a 10-gallon jug}
\end{array}
$$

Now we need to find the number of liters in a 10-gallon jug. There are 0.946 quarts per liter and 4 quarts per gallon. So we multiply the number of quarts in a gallon, 4, times the amount of quarts in a liter, 0.946.

$$
\begin{array}{r}
0.946 \quad \text{quarts per liter} \\
\times\quad 4 \quad \underline{\text{quarts in a gallon}} \\
3.784 \quad \text{liters in a gallon}
\end{array}
$$

This gives us the number of liters in one gallon. Since we're trying to find out how many liters are in a 10-gallon jug, we'll multiply 3.784 liters times 10.

$$
\begin{array}{r}
3.784 \quad \text{liters in a gallon} \\
\times\ 10 \quad \underline{\text{gallons}} \\
37.84 \quad \text{liters in a 10-gallon jug}
\end{array}
$$

Now that we know there are 37.84 liters and 1,280 ounces in a 10-gallon jug, we're ready to answer the questions.

a. If the team buys all the liquid butter at the convenience store, they will buy it in 20-ounce bottles. They need 1,280 ounces to fill that 10-gallon jug. To find out how many bottles they need, we must find out how many times 20 will go into 1,280.

$$20\overline{)1280}$$

We have to divide 20 into 128, since 20 will not go into 12. We can divide 128 by 20 six times.

$$\overset{6}{20\overline{)1280}}$$

We multiply 6 times 20.

$$6 \times 20 = 120$$

$$\overset{6}{20\overline{)1280}}\\\underline{120}$$

Then we subtract 120 from 128 and bring down the zero.

$$\overset{6}{20\overline{)1280}}\\\underline{120}\\80$$

We can divide 20 into 80 four times, leaving us with
no remainder.

$$
\begin{array}{r}
64 \\
20)\overline{1280} \\
\underline{120} \\
80 \\
\underline{80} \\
0
\end{array}
$$

The team will need 64 bottles exactly. Now, if each bottle costs
0.85, then the total price will be 64 × 0.85, or

$$
\begin{array}{r}
64 \\
\times\,.85
\end{array}
$$

We multiply 5 times all the digits on top, carrying
when needed.

$$
\begin{array}{r}
64 \\
\times\,.85 \\
\hline
320
\end{array}
$$

Before we multiply the 8 times all the digits in the top number,
we must put a zero in the units place of the second result.
The units place of each result must go beneath the bottom
digit we're multiplying by.

$$
\begin{array}{r}
64 \\
\times\,.85 \\
\hline
320 \\
0
\end{array}
$$

Now we just multiply 8 times both digits in the top number, carrying where its necessary.

$$\begin{array}{r} 64 \\ \times\ .85 \\ \hline 320 \\ \underline{5120} \end{array}$$

Our last step is to add the results and move the decimal point two places to the left in our product.

$$\begin{array}{r} 64 \\ \times\ .85 \\ \hline 320 \\ \underline{5120} \\ 54.40 \end{array}$$

Now we know it's going to cost the team $54.40 to buy all the liquid butter at the convenience store. We've finished the first question.

b. If the soccer team buys all the liquid butter at the grocery store, they will buy it in 1-liter bottles. We need 37.84 liters to fill that 10-gallon jug, so we divide 1, the amount of liters in each bottle, into 37.84. We know immediately that we need 37.84 bottles because anything divided by 1 is just itself. That means they will have to buy 38 bottles and use whatever is left over after they fill up the jug.

The team needs 38 bottles. The price of each bottle is $0.65. So we multiply the number of bottles, 38, times the price of each bottle, $0.65.

$$
\begin{array}{r}
38 \\
\times\ .65 \\
\hline
\end{array}
$$

We solve this problem like we did the last one. We multiply 5 times all the digits in the top number.

$$
\begin{array}{r}
38 \\
\times\ .65 \\
\hline
190 \\
\end{array}
$$

We place a zero in the units place of our second result, then multiply 6 times both digits in the top number.

$$
\begin{array}{r}
38 \\
\times\ .65 \\
\hline
190 \\
2280 \\
\hline
\end{array}
$$

Our last step is to add the results and place the decimal point two digit places from the right end of our product.

$$
\begin{array}{r}
38 \\
\times\ .65 \\
\hline
190 \\
2280 \\
\hline
24.70 \\
\end{array}
$$

It will cost $24.70 to buy all the liquid butter at the grocery store. Right on. One more question left.

c. Okay, now, we need to find our how much it will cost to buy $\frac{1}{3}$ at the convenience store and $\frac{2}{3}$ at the grocery store. Our first step is to find out how many ounces the team needs from each store. We start by figuring out the number of ounces they'll need at the convenience store. Remember, they need a total of 1,280 ounces. So we'll multiply $\frac{1}{3}$ times this amount.

$$\frac{1}{3} \times \frac{1280}{1} = \frac{1280}{3}$$

Now we divide our denominator into our numerator to find the number of ounces we need.

$$1280 \div 3 = 426\frac{2}{3}$$

Next, we find the number of 20-ounce bottles needed. To do this, we must divide 20, the number of ounces in each bottle, into $426\frac{2}{3}$.

This will be easier to do with fractions than long division. Remember, when dividing fractions, you flip the fraction that is the divisor and then multiply the two fractions together. Instead of dividing by 20, we will multiply by $\frac{1}{20}$, like this:

$$\frac{1280}{3} \times \frac{1}{20} = \frac{1280}{60}$$

We divide the denominator into the numerator to find that we need a little more than 21 bottles, which means we need 22. Okay! We'll have to buy 22 bottles at $0.85, pour most of it in the jug, and put the rest on our ice cream. So, let's figure out what this will cost. We multiply the 22 bottles times the cost per

bottle. Here's what the problem looks like when we work it out.

```
   .85
 ×22
  170
 1700
18.70
```

Natch. It'll cost $18.70 to buy the liquid butter at the convenience store.

We still need to buy $\frac{2}{3}$ of 37.84 liters at the grocery store. Remember, 37.84 liters is how many liters it takes to fill a gallon, and we're buying $\frac{2}{3}$ of our gallon at the grocery store. First we find the number of liters we need. To simplify things, we'll convert $\frac{2}{3}$ into a decimal. Dividing the denominator, 3, into the numerator, 2, we get .67. Then, we multiply .67 times 37.84, to find the amount of butter to buy at the grocery store.

```
  37.84
 × .67
 26488
227040
25.3528
```

This is the amount of liters needed. Now we have to find the number of liter bottles they need. This is easy, since there's 1 liter in each bottle, because all we have to do is round up to the nearest whole number. So, 25.3528 rounds up to 26 bottles.

Okay! The team will have to buy 26 bottles at $0.65. Let's figure out what this will cost:

$$
\begin{array}{r}
26 \\
\times\ .65 \\
\hline
130 \\
\underline{156} \\
16.90
\end{array}
$$

It's going to cost $16.90 for the liquid butter at the grocery store. We're almost done with this problem! All we have to do now is add up the $18.70 we'll spend at the convenience store and the $16.90 we'll spend at the grocery store.

$$
\begin{array}{r}
18.70 \\
+\ \underline{16.90} \\
35.60
\end{array}
$$

So, by buying the liquid butter in two places, the soccer team ends up paying $35.60.

Okay, do these answers make sense? Let's check! The arithmetic is correct in all the parts of this problem and we've remembered to use equivalent units for all parts. Everything checks out just fine so far.

What about our logic? We know that at the convenience store we're buying a smaller amount at a higher price than at the grocery store. The bill should be higher at the convenience store than the grocery store, and it is. That fits! Splendid.

Now, if we buy part of our whole gallon at a high price and part at a low price, the final bill should be less than if we

bought all the liquid butter at the convenience store, but more than if we bought all the liquid butter at the grocery store. And you betcha, it is. It all fits. All is well! We're done!!!! Since that problem was such fun, we thought you'd like to do some more, so we added some into the fun stuff section. You can thank us later.

STRESS RELIEF

Basic Searcho

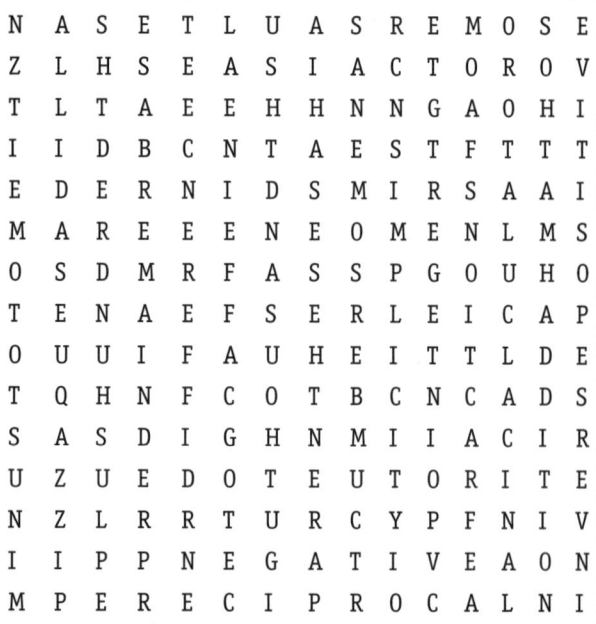

```
N  A  S  E  T  L  U  A  S  R  E  M  O  S  E
Z  L  H  S  E  A  S  I  A  C  T  O  R  O  V
T  L  T  A  E  E  H  H  N  N  G  A  O  H  I
I  I  D  B  C  N  T  A  E  S  T  F  T  T  T
E  D  E  R  N  I  D  S  M  I  R  S  A  A  I
M  A  R  E  E  E  N  E  O  M  E  N  L  M  S
O  S  D  M  R  F  A  S  S  P  G  O  U  H  O
T  E  N  A  E  F  S  E  R  L  E  I  C  A  P
O  U  U  I  F  A  U  H  E  I  T  T  L  D  E
T  Q  H  N  F  C  O  T  B  C  N  C  A  D  S
S  A  S  D  I  G  H  N  M  I  I  A  C  I  R
U  Z  U  E  D  O  T  E  U  T  O  R  I  T  E
N  Z  L  R  R  T  U  R  C  Y  P  F  N  I  V
I  I  P  P  N  E  G  A  T  I  V  E  A  O  N
M  P  E  R  E  C  I  P  R  O  C  A  L  N  I
```

ADDITION, CAFFEINE, CALCULATOR, CUMBERSOME, DIFFERENCE, ELEVATOR, FRACTION, HUNDREDTHS, IMPROPER, JUJITSU, PARENTHESES, PERCENT, POINT, QUESADILLA, RATIO, RECIPROCAL, REMAINDER, SIMPLICITY, SOMERSAULT, SQUAREDANCE ,STOTOMEITZ, THOUSANDTHS, WHOLE

Basic Crosso

The Zany World of Basic Math is great! Great! Simply great!

– Hossiferson the Great

MATH STUFF & 80's ROCK GROUPS

ACROSS

6. grassy area of baseball field
10. what's left over in a division problem
14. glam rock band or arsenic
15. do this if digit place sum 9 or greater
17. number to left of division sign
18. place to left of decimal point
19. answer to multiplication problem
21. addition symbol
22. number of times multiply base by
27. move decimal point _____ multiplying
28. band with comic book video
29. notation useful for large numbers
32. group who made Pyromania
33. shape of burger or an approximate guess

DOWN

1. do this when top # is smaller than bottom #
2. digit place of #1 in number 15
3. sang "Centerfold"
4. exponent multiplies this times itself
5. Boy George lead singer of
7. band members later formed Power Station
8. Frankie goes to _____
9. # of digit places move to multiply by %
11. _____ midnight runners
12. bottom number of fraction
13. you flip a fraction over to find its _____
14. "Don't Stand So Close To Me"
16. numbers 0 to 9 are examples of this
20. digit place to right of decimal point
21. numbers larger than zero are this
23. part fraction, part whole number
24. this symbol "%" means what
25. move decimal point _____ dividing
26. name for relationship between qualities
30. human bee in video or what Chevys are
31. to find a sum you have to do this

147

Magic Squares

Magic squares are sets of integers that can be arranged in a square formation, so that the sum of each row, column, and main diagonal (top left to bottom right and bottom left to top right) is the same number. These squares have a long history. Many cultures have studied them since ancient times. Some think that these squares can bring you luck. What were they thinking? It's not like they'd found a rabbit's foot or anything.

The square that follows is an example of a magic square of order three. Order refers to the number of rows or columns in the square.

2	7	6
9	5	1
4	3	8

Notice that the sum of each row, column, and main diagonal equals 15.

Now it's your turn. Fill in the missing numbers in the following magic squares:

	35	30
	25	5
20		

7		1	14
2	13		11
	3	10	5
9	6	15	4

	2	3	13
5		10	8
9	7		12
4	14		1

Word Problems

1. Ima Big Car Dealership is having a sale on last year's models. Ima Big, president of the dealership, is advertising a rebate of $1,750 on all cars that cost more than $17,500.

   a. What is the cost, after the rebate has been applied, of a car with a sticker price of $21,995?

   b. What percent of the sticker price is the rebate for a car with a sticker price of $17,500?

   c. What percent of the sticker price is the rebate for a car with a sticker price of $25,000?

2. Ima Small Car Dealership sells the same makes of cars as Ima Big Car Dealership. It decides to offer a rebate of 10% off the sticker price of all cars in stock to try to steal customers from Ima Big Car Dealership's sale.

   a. What is the cost, after the rebate has been applied, of a car with a sticker price of $21,995?

b. If a car's final cost is $20,000, what did it cost before the 10% discount was applied?

3. Karen has decided to purchase a car from Ima Small Car Dealership for a final price of $30,000. She wants to pay $3,499 as a down payment and then pay off the rest in monthly payments.

   a. How much is left for her to pay after the down payment?

   b. If she pays $4,000 in interest and can take 60 months to pay off the principal and interest, how much is her monthly payment?

4. Mitch, Karen's husband, earns $20.65 per hour for each of the first 40 hours he works in a week. He earns 150% of $20.65 per hour for each additional hour he works during that week.

   a. How much will Mitch earn in a year if he works 40 hours per week for 50 weeks?

   b. How much will Mitch earn in a week if he works 46.5 hours during that week?

   c. If Mitch's federal taxes are 15% of his gross pay, his social security and Medicare taxes are 7.65% of his gross pay, and his state and local taxes are 6% of his gross pay, how much can he take home after a week if his gross pay is $1,155.00?

5. Mitch and Karen are driving around in Karen's new car. They stop to do some grocery shopping and make one long-distance call from a pay phone.

   a. If Karen buys $\frac{3}{4}$ of a pound of turkey at $4.99 a pound, half a pound of Swiss cheese at $3.69 per pound, two tomatoes at 3 for $1.20, and 3 Kaiser rolls for $3.00 per dozen, what is her bill?

   b. If Mitch makes a 15-minute call and the phone company charges $0.65 for the first minute and $0.36 for each additional minute, how much will his call cost?

ANSWERS

Quiz 1

1. Add the following integers.

    a. 35 + 6

$$
\begin{array}{r}
{\scriptstyle 1} \\
35 \\
+\ \ 6 \\
\hline
41
\end{array}
$$

    b. 245 + 397

$$
\begin{array}{r}
{\scriptstyle 1\,1} \\
245 \\
+\ 397 \\
\hline
642
\end{array}
$$

2. Subtract the following integers.

    a. 35 − 6

$$
\begin{array}{r}
{\scriptstyle 2\,15} \\
\cancel{35} \\
-\ \ 6 \\
\hline
29
\end{array}
$$

153

b. $100 - 38$

$$
\begin{array}{r}
\overset{9\ 10}{1\cancel{0}\cancel{0}} \\
-\ \ 38 \\
\hline
62
\end{array}
$$

Another day,
another dollar;
14 hours on
snowshoes
and wish I
had a pie.

– From a Maine
trapper's diary

3. Multiply the following integers.

a. 41×27

$$
\begin{array}{r}
41 \\
\times\ 27 \\
\hline
287 \\
820 \\
\hline
1107
\end{array}
$$

b. 135×89

$$
\begin{array}{r}
135 \\
\times\ \ 89 \\
\hline
1215 \\
10800 \\
\hline
12{,}015
\end{array}
$$

4. Simplify the following by removing the exponents.

a. 2^4

$$2 \times 2 \times 2 \times 2 = 16$$

b. 16^2

$$16 \times 16 = 256$$

ANSWERS

5. Divide the following integers.

    a. $512 \div 8$

$$
\begin{array}{r}
64 \\
8\overline{)512} \\
\underline{48} \\
32 \\
\underline{32} \\
0
\end{array}
$$

    b. $1690 \div 13$

$$
\begin{array}{r}
130 \\
13\overline{)1690} \\
\underline{13} \\
39 \\
\underline{39} \\
0
\end{array}
$$

6. Solve the following strings of integers, paying careful attention to the order of operations.

    a. $2 + 3 \div 1 - 2$

$$
\begin{aligned}
2 + 3 \div 1 - 2 &= 2 + 3 - 2 \\
&= 5 - 2 \\
&= 3
\end{aligned}
$$

    b. $2 \times (14 - 9) - 1$

$$
\begin{aligned}
2 \times (14 - 9) - 1 &= 2 \times 5 - 1 \\
&= 10 - 1 \\
&= 9
\end{aligned}
$$

7. Round off the following integers to the nearest hundred.

  a. 1,545

  1500

  b. 291

  300

8. Write the following integers in scientific notation.

  a. 1,000,000

  1×10^6

  b. 3,000

  3×10^3

Quiz 2

1. What place is occupied by the numeral 5 in the decimal numbers below?

  a. 5,046.32

  thousands (4th place to the left of the decimal point)

  b. 997.95

  hundredths (2nd place to the right of the decimal point)

2. Perform the following additions and subtractions of decimal numbers.

a. 123.5 + 234.6 + 98.09

$$
\begin{array}{r}
^{1\,1\,1} \\
123.5 \\
234.6 \\
+\ \underline{98.09} \\
456.19
\end{array}
$$

b. 9,780.65 − (234.5 − 198.2)

We do the subtraction in parentheses first:

$$
\begin{array}{r}
^{1\,12\,14} \\
2\cancel{34}.5 \\
-\underline{198.2} \\
36.3
\end{array}
$$

Now we have:

$$
\begin{array}{r}
^{7\,10} \\
978\cancel{0}.65 \\
-\ \underline{36.30} \\
9744.35
\end{array}
$$

Done!

c. 123.5 + (234.6 − 98.1)

We do the subtraction in the parentheses first:

$$
\begin{array}{r}
{}^{1\ 12\ 14} \\
\cancel{234}.6 \\
-\ \underline{\ 98.1} \\
136.5
\end{array}
$$

And we're left with:

$$
\begin{array}{r}
{}^{1\ 1} \\
123.5 \\
+\ \underline{136.5} \\
260.0
\end{array}
$$

3. Perform the following multiplications and divisions of decimal numbers.

   a. 54.2×19.1

$$
\begin{array}{r}
54.2 \\
\times\ \underline{19.1} \\
542 \\
48780 \\
\underline{54200} \\
1035.22
\end{array}
$$

   b. $61.8 \div 2 \times 1.5$

First we'll do the division:

```
      30.9
   2)61.8
      6
      ‾‾
      01
      00
      ‾‾
      18
      18
      ‾‾
       0
```

Now we multiply our quotient by 1.5.

```
      30.9
   ×  1.5
   ‾‾‾‾‾‾
      1545
      3090
   ‾‾‾‾‾‾
     46.35
```

c. $64.5 \div 0.5 \times 9 \div 4.5$

We'll do the division problem on the left first.

```
      129
   5)645
      5
      ‾‾
      14
      10
      ‾‾
      45
      45
      ‾‾
       0
```

Now we multiply by 9.

$$\begin{array}{r} {\scriptstyle 2\,8} \\ 129 \\ \times \quad 9 \\ \hline 1161 \end{array}$$

Finally, we divide by 4.5.

$$\begin{array}{r} 258 \\ 45\overline{)11610} \\ \underline{90} \\ 261 \\ \underline{225} \\ 360 \\ \underline{360} \\ 0 \end{array}$$

4. Round off these decimal numbers to the nearest hundredth.

　a. 35.045666667

$$35.05$$

　b. 9.9292929292

$$9.93$$

5. Write these decimal numbers in scientific notation.

　a. 0.00000789

$$7.89 \times 10^{-6}$$

　b. 12.0005

$$1.20005 \times 10^{1}$$

Quiz 3

1. What are the numerators and the denominators of these fractions?

   a. $\dfrac{12}{13}$

   The numerator is 12. The denominator is 13.

   b. $\dfrac{5}{7}$

   The numerator is 5. The denominator is 7.

2. Add or subtract the following fractions.

   a. $\dfrac{1}{2} + \dfrac{3}{5}$

$$\frac{1}{2} + \frac{3}{5} = \frac{1 \times 5}{2 \times 5} + \frac{3 \times 2}{5 \times 2}$$

$$= \frac{5}{10} + \frac{6}{10}$$

$$= \frac{5 + 6}{10}$$

$$= \frac{11}{10}$$

$$= 1\frac{1}{10}$$

I have to be deliberately vague to be correct.

–Jennie Halfant

b. $\dfrac{8}{9} - \dfrac{1}{3} + \dfrac{5}{6}$

$$\frac{8}{9} - \frac{1}{3} + \frac{5}{6} = \frac{8 \times 2}{9 \times 2} - \frac{1 \times 6}{3 \times 6} + \frac{5 \times 3}{6 \times 3}$$

$$= \frac{16}{18} - \frac{6}{18} + \frac{15}{18}$$

$$= \frac{16 - 6 + 15}{18}$$

$$= \frac{25}{18}$$

$$= 1\frac{7}{18}$$

3. Multiply or divide these fractions.

a. $\dfrac{1}{2} \times \dfrac{3}{4}$

$$\frac{1}{2} \times \frac{3}{4} = \frac{1 \times 3}{2 \times 4}$$

$$= \frac{3}{8}$$

b. $\dfrac{5}{7} \div \dfrac{25}{70}$

$$\dfrac{5}{7} \div \dfrac{25}{70} = \dfrac{5}{7} \times \dfrac{70}{25}$$

$$= \dfrac{5 \times 70}{7 \times 25}$$

$$= \dfrac{350}{175}$$

$$= \dfrac{2}{1}$$

$$= 2$$

4. Perform the indicated operations on these mixed numbers.

a. $1\dfrac{1}{3} - \dfrac{5}{6}$

$$1\dfrac{1}{3} - \dfrac{5}{6} = \dfrac{3}{3} + \dfrac{1}{3} - \dfrac{5}{6}$$

$$= \dfrac{4}{3} - \dfrac{5}{6}$$

$$= \dfrac{4 \times 2}{3 \times 2} - \dfrac{5}{6}$$

$$= \dfrac{8}{6} - \dfrac{5}{6}$$

$$= \dfrac{8 - 5}{6}$$

$$= \dfrac{3}{6}$$

$$= \dfrac{1}{2}$$

b. $1\frac{3}{7} \times 5\frac{3}{4}$

$$1\frac{3}{7} \times 5\frac{3}{4} = \frac{10}{7} \times \frac{23}{4}$$

$$= \frac{10 \times 23}{7 \times 4}$$

$$= \frac{230}{28}$$

$$= \frac{115}{14}$$

$$= 8\frac{3}{14}$$

Quiz 4

1. Find the following ratios:

a. Frank looks in his freezer and sees that he has 100 fudgecicles, 40 grape popsicles, and 60 cases of motor oil. As you can tell, Frank's got a big freezer. What is the ratio of the objects in Frank's freezer? (reduce to lowest possible ratio)

Solution:

Remember that ratios are similar to fractions. Each of the quantities represents a part present, and the total of the quantities is the whole. We'll start off by adding up the quantities.

100 fudgecicles + 40 grape popsicles + 60 cases of motor oil = 200 total items

Now we can make each of our quantities a fraction by making the part present the numerator and the total items the denominator.

$$\frac{100}{200} \qquad \frac{40}{200} \qquad \frac{60}{200}$$

Notice that each of these fractions can be reduced by 20. After reducing each fraction, we have the following fractions:

$$\frac{5}{10} \qquad \frac{2}{10} \qquad \frac{3}{10}$$

This means our lowest possible ratio is 5:2:3.

b. Marsha supervises a fleet of cars that her company rents. If they have 6 stretch Cadillacs and 4 Corvettes made by General Motors, and 36 Lincoln Town Cars and 24 Broncos made by Ford Motor Company, what is the ratio in the fleet of vehicles made by GM to vehicles made by Ford?

Solution:

There are 6 + 4 = 10 GM vehicles and 36 + 24 = 60 Ford vehicles. This means that the ratio is 10:60. We can simplify this ratio to 1:6.

2. Convert the following numbers into percents.

a. $\dfrac{1}{4}$

We divide the denominator into the numerator.

$$
\begin{array}{r}
0.25 \\
4\overline{)1.00} \\
\underline{0} \\
10 \\
\underline{8} \\
20 \\
\underline{20} \\
0
\end{array}
$$

Now we move the decimal point 2 places to the right and insert a percent symbol.

$$0.25 \rightarrow 25\%$$

b. 2

2 is the same as 2.00 or 200%

Basic Searcho

Basic Crosso

I'm going to
name my son
Flirpdü!

– Joe Braband

Magic Squares

Okay, for the first one, we are starting with this:

10	35	30
45	25	5
20	15	40

We'll start with what we know. We know that each row, column, and major diagonal added up to the same amount.
So, we see that we have one complete major diagonal. We add those numbers together.

$$20 + 25 + 30 = 75$$

This means that every column, row and diagonal in this triangle equals 75.

Next, we'll look at the second column, since it's only missing one number. We know that are two given numbers (35 and 25) plus another number, must equal 75. Our next step is to add 35 and 25 together.

$$
\begin{array}{r}
35 \\
+\ 25 \\
\hline
60
\end{array}
$$

Okay, now we know the second column in the magic square equals 60, and we want it to equal 75. We have to find out what number can be added to 60 to get 75. This is a subtraction problem, so we subtract 60 from 75, like this:

$$
\begin{array}{r}
75 \\
-\ 60 \\
\hline
15
\end{array}
$$

The missing number in the second column is 15. We can write that in the magic square.

Let's find the missing number in the third column of the magic square. We have 30 and 5 in that column already, which adds up to:

$$30 + 5 = 35$$

So, like we just did, we have to subtract 35 from 75 to find the missing number:

$$
\begin{array}{r}
75 \\
- \ 35 \\
\hline
40
\end{array}
$$

Abracadabra. The missing number is 40 and we can write that in the magic square.

Did you notice that we used the same method to find each of the first two missing numbers? Once we found the magic number, the one that is the sum of the numbers in each row, column, and major diagonal, we subtracted the sum of the numbers we knew from the magic number. Here's what our magic square looks like once we find all the numbers.

You can follow the same process as we did with the previous square to find the other two. Here's the answer to the second magic square.

7	12	1	14
2	13	8	11
16	3	10	5
9	6	15	4

Our last magic square looks like this.

16	2	3	13
5	11	10	8
9	7	6	12
4	14	15	1

Word Problems

1. Ima Big Car Dealership is having a sale for last year's models to move them out the door and into your driveways. Ima Big, president of the dealership, is advertising a rebate of $1,750 on all cars that cost more than $17,500.

    a. What is the cost, after the rebate has been applied, of a car with a sticker price of $21,995?

    b. What percent of the sticker price is the rebate for a car with a sticker price of $17,500?

    c. What percent of the sticker price is the rebate for a car with a sticker price of $25,000?

Solution:

Okay, first we need to figure out how to use this rebate. Ima Big is taking $1,750 away from the price of every car that costs more than $17,500. This means we first check to see if the car is sell-

ing for more than $17,500. If it is, we subtract $1,750 off the sticker price to find the final price. If it isn't, we can't apply the rebate and the price doesn't change.

a. For a car with a sticker price of $21,995, the price is more than $17,500, so we can apply the rebate. We find the final price by subtracting $1,750 from $21,995.

$$\begin{array}{r} 21995 \\ -\ 1750 \\ \hline 20,245 \end{array}$$

The car will cost $20,245 after the rebate.

b. For this part of the problem, remember that the rebate is ONLY offered on cars with sticker prices OVER $17,500. A car with a sticker price of exactly $17,500 gets no rebate (or, you could say it gets a rebate of $0). This means that the percent is also 0%. Too bad.

c. We'll start by checking to see if this car is eligible for a rebate. We can see that the car is eligible, since $25,000 is more than $17,500. We're trying to find out what percent the rebate ($1,750) is of 25,000. Essentially, we're calculating what part $1,750 is of the whole ($25,000). We'll put these two amounts into a fraction with $1,750 as the numerator (the part present) and $25,000 as the denominator (the whole).

$$\frac{1750}{25000}$$

Remember, when you have a fraction and you're trying to find a percent you first divide the denominator into the numerator. Our first step, then, in figuring out the percent, is to divide the sticker price, 25,000, into the amount of the rebate, 1,750.

$$25000\overline{)1750}$$

Notice that 1,750 is less than 25,000, which tells us our quotient is going to be less than one. Whenever your dividend is smaller than your divisor, you can add zeroes to the end of your dividend until it is divisible. Here, we need to add 2 zeroes to make our dividend larger than the divisor. But if you add zeroes to the end of your product, you have to place a decimal point before these added zeroes. That way, you still haven't changed the value of the dividend.

$$25000\overline{)1750.00}$$

Now we can just divide the problem out like we normally would. In numbers this large, you have to solve the division problem through trial and error. Here's what the division looks like when it's worked out.

$$
\begin{array}{r}
.07 \\
25000\overline{)1750.00} \\
\underline{1750\ 00} \\
0
\end{array}
$$

Once we have a decimal, we just multiply it by 100, or move the decimal point 2 places to the right.

$$.07 \rightarrow 7\%$$

Our final answer is 7% (don't forget the percent sign).

2. Ima Small Car Dealership sells the same makes of cars as Ima Big Car Dealership. It decides to offer a rebate of 10% off the sticker price of all cars in stock to try to steal customers from Ima Big Car Dealership's sale.

    a. What is the cost, after the rebate has been applied, of a car with a sticker price of $21,995?

    b. If a car's final cost is $20,000, what did it cost before the 10% discount was applied?

Solution:

Okay, reading this problem tells us there's a rebate on every car, so this time we don't have to check to see which cars are eligible and which aren't.

a. All we have to do to calculate the price of the car after the rebate is multiply the sticker price by 10% and then subtract the rebate from the sticker price. The rebate on this car is found by multiplying 10% by $21,995. Our first step is to make 10% a decimal. To do this, we move the decimal point 2 places to the LEFT. After moving the decimal point, we get 0.1. We'll multiply this number times 21, 995 to find the discount.

$$
\begin{array}{r}
21995 \\
\times \underline{\quad .10} \\
2{,}199.5
\end{array}
$$

All we did is move the decimal point one place to the left. Now we have the amount of the rebate. Let's subtract this amount from the sticker price of $21,995 to find out what the car cost before the rebate. Don't forget to the keep the decimal points lined up.

$$
\begin{array}{r}
2{,}1995.00 \\
- \underline{2{,}199.50} \\
19{,}795.50
\end{array}
$$

The final price is $19,795.50.

b. We need to find the price of a car before the rebate. We know from the price after the rebate that the rebate is 10% of the sticker price. But we don't know the amount of the rebate. What do we do? Well, since we know the rebate is 10% of the sticker price, we ALSO know the final price is 90% of the

sticker price. So, if we divide the final price by 90%, we can find the sticker price and not have to mess around with figuring out the rebate. Like our last problem, we want to make 90% a decimal. We just move the decimal point 2 places to the left.

$$90\% \rightarrow 0.90 \text{ or simply } .9$$

Now we just divide the price of the car after the rebate, $20,000, by .9. Remember, when we have a decimal in the divisor, we must move the decimal point until we have a whole number. Then we move the decimal point in the dividend the same number of places.

```
        22222.22
    9)200000.00
      18
       20
       18
        20
        18
         20
         18
          20
          18
           20
           18
            20
            18
             2
```

So, rounding off to the nearest penny, the sticker price is $22,222.22. You can check this by finding the amount of the rebate and subtracting the rebate from the sticker price, just like we did before.

3. Karen has decided to purchase a car from Ima Small Car Dealership for a final price of $30,000. She wants to pay $3,499 as a down payment and then pay off the rest in monthly payments.

   a. How much will she have to pay in monthly payments?

   b. If she pays $4,000 in interest and can take 60 months to pay off the principal and interest, how much is her monthly payment?

Solution:

a. Her monthly payments are the final price, minus the down payment. So we subtract.

$$\begin{array}{r} {\scriptstyle 2\ 9\ \ 9\ 9\ 10} \\ \cancel{3}\cancel{0},\cancel{0}\cancel{0}\cancel{0} \\ -\ \underline{3,499} \\ 26,501 \end{array}$$

The principal amount she will have to pay is $26,501.00 (notice that you have to borrow to solve this problem).

ANSWERS

b. Ima will have to pay the principal of $26,501.00 plus the $4,000.00 in interest over 60 months. Her monthly payment will be $\frac{1}{60}$ of the sum of $26,501.00 plus $4,000.00. First we're going to have to add the principal and interest together.

$$
\begin{array}{r}
26501 \\
+\ \underline{4000} \\
30,501
\end{array}
$$

We know that Karen will have to pay $30,501 in principal and interest over 60 months. Our next step is to divide 60 into 30,501.00 to find the monthly payment.

$$
\begin{array}{r}
508.35 \\
60\overline{)30501.00} \\
\underline{300} \\
50 \\
\underline{0} \\
501 \\
\underline{480} \\
210 \\
\underline{180} \\
300 \\
\underline{300} \\
0
\end{array}
$$

Mitch has to pony up $508.35 for 60 months to pay for this car.

STUDY SIDEKICK

4. Mitch, Karen's husband, earns $20.65 per hour for each of the first 40 hours he works in a week and earns 150% of $20.65 per hour, for each additional hour he works during that week.

   a. How much will Mitch earn in a year if he works 40 hours per week for 50 weeks?

   b. How much will Mitch earn in a week if he works 46.5 hours during that week?

   c. If Mitch's federal taxes are 15% of his gross pay, his social security and Medicare taxes are 7.65% of his gross pay, and his state and local taxes are 6% of his gross pay, how much can he take home after a week when his gross pay is $1,155.00?

Solution:

In this problem, you have to see how many regular hours are worked per week and how many of the hours he worked are overtime hours.

a. Mitch does get overtime pay, but only for hours worked over 40 hours in each week. So, if he works 40 hours per week at $20.65 per hour for 50 weeks, we have a series of multiplications to do. First, let's find out how much Mitch makes in a 40-hour work week. We'll multiply the 40 hours Mitch works each week times his hourly rate, $20.65.

$$
\begin{array}{r}
20.65 \\
\times \quad 40 \\
\end{array}
$$

To simplify this multiplication problem, we'll multiply 4 times 20.65 and then add the zero to the end of our product.

$$
\begin{array}{r}
20.65 \\
\times 4 \\
\hline
82.60
\end{array}
$$

Now we add the zero to get $826.00. Okay, we now see that Mitch grosses $826.00 for each 40 hour work week. We know he works for 50 weeks. So we'll multiply his weekly gross pay, $826.00, times the number of weeks worked, 50. Like in the previous multiplication problem, we'll multiply 826 times 5 and add the zero at the end of the product.

$$
\begin{array}{r}
826 \\
\times 5 \\
\hline
4{,}130
\end{array}
$$

Now we add the zero to get $41,300. This tells us that Mitch makes $41,300.00 if he works 40 hours per week for 50 weeks at $20.65 per hour.

b. For this part of the problem, Mitch is making overtime pay for some of his hours. Therefore we have to split the hours into two parts, regular and overtime. Then, we have to figure out just what his overtime pay is per hour, to be able to calculate what his gross pay is for the week. First, we split his hours. Since 40 hours will be his regular pay, we need to find the difference between 46.5 and 40.

$$
\begin{array}{r}
46.5 \\
-\ \underline{40.0} \\
6.5
\end{array}
$$

Now we know how many regular hours and overtime hours Mitch has worked. Mitch gets paid 40 hours at $20.65 per hour and 6.5 hours at 150% of $20.65 per hour. From the previous problem, we know that Mitch gets paid $826.00 per week if he works forty hours. We need to find out how much Mitch gets paid for these 6.5 hours of overtime. Let's change 150% to a decimal first. We move our decimal point 2 places to the left.

$$150\% \rightarrow 1.5$$

Next, we multiply 1.5 times his hourly rate, $20.65, to find out how much Mitch gets paid for each overtime hour he works.

$$
\begin{array}{r}
20.65 \\
\times\ \underline{1.5} \\
10325 \\
\underline{20650} \\
30.975
\end{array}
$$

His overtime rate is $30.975 per hour. Now, Mitch works 6.5 hours at that rate, so his overtime pay for the week will be $30.975 times 6.5, like this:

$$
\begin{array}{r}
30.975 \\
\times \quad 6.5 \\
\hline
154875 \\
1858500 \\
\hline
201.3375
\end{array}
$$

Okay, we're almost done. Mitch made $201.34 (rounding to the nearest penny) for his overtime hours. To find his total pay for 46.5 hours worked we must add his total pay for regular hours to his overtime pay.

$$
\begin{array}{r}
826.00 \\
+ \quad 201.34 \\
\hline
1027.34
\end{array}
$$

At last! Mitch made a gross total of $1,027.34 for a week when he worked 46.5 hours. Bravo.

c. To find how much of Mitch's gross pay will be withheld for these taxes, we'll have to multiply his gross pay, $1,155.00, times each of these percentages. To figure out his federal taxes, we have to find out what 15% of $1,155.00 is. This is a simple multiplication problem. We just switch 15% to 0.15 then multiply it by 1,155.

$$
\begin{array}{r}
1155 \\
\times\ .15 \\
\hline
5775 \\
\underline{11550} \\
172.25
\end{array}
$$

Okay, Mitch has $172.25 withheld for federal taxes. Now, how much will be withheld for social security and Medicare taxes? This is 7.65%, or .0765, of $1,155.00.

$$
\begin{array}{r}
1155 \\
\times\ .0765 \\
\hline
5775 \\
69300 \\
\underline{808500} \\
88.3575
\end{array}
$$

Rounding to the nearest penny, Mitch's social security and Medicare withholding is $88.36.

Now, Mitch still has to have his state and local taxes withheld from his gross pay. This will be 6% of $1,155.00. Here we go again. We change 6% to 0.06. Then multiply it by $1,155.

$$
\begin{array}{r}
1155 \\
\times \ .06 \\
\hline
69.30
\end{array}
$$

So, Mitch has $69.30 withheld from his pay for his state and local taxes.

Now we have found that Mitch has had $172.25, $88.36, and $69.30 withheld from his gross pay of $1,155.00. Our last step is to subtract these withheld taxes from his gross pay. The easiest way to do this is to add the withheld taxes together, then subtract their sum from Mitch's gross pay.

$$
\begin{array}{r}
172.25 \\
88.36 \\
+ \ 69.30 \\
\hline
329.91
\end{array}
$$

We're at the end of the problem. All we have to do is subtract $329.91 in total withheld taxes from $1,155.00 in gross pay:

$$
\begin{array}{r}
1155.00 \\
- \ 329.91 \\
\hline
825.09
\end{array}
$$

STUDY SIDEKICK

Cool. We're done. Mitch will take home $825.09.

5. Mitch and Karen are driving around in Karen's new car. They stop to do some grocery-shopping and make one long-distance call from a pay phone.

   a. If Karen buys $\frac{3}{4}$ of a pound of honey turkey at $4.99 a pound, $\frac{1}{2}$ a pound of Swiss cheese at $3.69 per pound, 2 tomatoes at 3 for $1.20, and 3 Kaiser rolls at $3.00 per dozen, what is her bill?

   b. If Izzie makes a 15 minute call and the phone company charges $0.65 for the first minute and $0.36 for each additional minute, how much will his call cost?

Solution:

a. Here we have to break this problem down into each item bought, and then just add up the costs of the items. So, here goes.

Turkey Part: $\frac{3}{4}$ of a pound at $4.99 per pound doesn't require any change of units. All we have to do is convert $\frac{3}{4}$ into a decimal and then multiply this by $4.99. First we'll convert our fraction into a decimal. We can divide the denominator into the numerator.

$$
\begin{array}{r}
.75 \\
4\overline{)3.00} \\
\underline{28} \\
20 \\
\underline{20} \\
0
\end{array}
$$

Now all we've got to do is multiply 0.75 times the price per pound.

$$
\begin{array}{r}
4.99 \\
\times\ \underline{.75} \\
2495 \\
\underline{34930} \\
3.7425
\end{array}
$$

Okay. $3.74 for the turkey. On to the cheese: $\frac{1}{2}$ lb is the same as 0.5 lb, so

$$
\begin{array}{r}
3.69 \\
\times\ \underline{.5} \\
1.845
\end{array}
$$

The cheese costs $1.85. On to the tomatoes. First we check out how much each tomato costs. They're 3 for $1.20. We divide 3 into $1.20 to find the cost of each one.

$$
\begin{array}{r}
.40 \\
3\overline{)1.20} \\
\underline{12} \\
00
\end{array}
$$

Each tomato cost $.40. Karen bought 2, so we multiply 2 by $.40.

$$
\begin{array}{r}
.40 \\
\underline{2} \\
.80
\end{array}
$$

The two tomatoes cost $0.80. Now, the Kaiser rolls are $3.00 per dozen and Karen buys 3 of them. 3 out of a dozen is the same as $\frac{3}{12}$. We can reduce this fraction to $\frac{1}{4}$. So Karen is buying $\frac{1}{4}$ dozen at $3.00 per dozen. We multiply $\frac{1}{4}$ times 3.

$$
\frac{1}{4} \times \frac{3}{1} = \frac{3}{4}
$$

We need to put this fraction back into decimal form. Remember we already converted this fraction once. We know that $\frac{3}{4}$ is the same thing as 0.75. The rolls cost her $0.75. Now, to find her total bill, we just add up the costs of all the items.

$$
\begin{array}{r}
3.74 \\
1.84 \\
0.80 \\
+\ \underline{0.75} \\
7.13
\end{array}
$$

We know that Karen's bill is $7.13.

b. Our first step in this problem is to separate the number of minutes Mitch was on the phone by their per-minute rates. We know that the first minute is $0.65. The first minute cost Mitch $0.65. Now we need to find out how much the rest of the minutes cost him. First, we'll find out how many additional minutes Mitch was on the phone. He was on the phone for 15 minutes. We subtract the first minute from this amount, since it has a different per minute rate. Mitch spoke for 14 minutes at the additional minute rate of $0.36. We multiply these amounts together.

> I can resist anything but temptation.
>
> – Oscar Wilde

$$
\begin{array}{r}
14 \\
\times\ .36 \\
\hline
84 \\
420 \\
\hline
5.04
\end{array}
$$

Now we just add the amount of money the first minute cost him to the total amount of money the 14 additional minutes cost.

$$
\begin{array}{r}
5.04 \\
+\ .65 \\
\hline
5.69
\end{array}
$$

Mitch's call cost him $5.69.

Practice Exam Answers

1. Name the position of the 7 in each of the numerals below:

    a. 0.736   7 is in the tenths place

    b. 7,562.65   7 is in the thousands place

    c. 137   7 is in the units place

    d. 63.467   7 is in the thousandths place

    e. 7,530,000   7 is in the millions place

    f. 0.00007   7 is in the hundred-thousandths place

2. Add the following integers:

    a. 729 + 1,036

$$\begin{array}{r} 1 \\ 729 \\ +\,1036 \\ \hline 1765 \end{array}$$

    b. 4,598 + 54 + 366

$$\begin{array}{r} 121 \\ 4598 \\ 54 \\ +\,366 \\ \hline 5018 \end{array}$$

c. 34 + 256,999 + 56 + 4 + 1101

```
        1 1 2
         34
     256999
         56
          4
   +   1101
     258194
```

d. 123 + 321 + 4,444

```
        123
        321
   +   4444
       4888
```

e. 46,789 + 56,301 + 67,943 + 11,009

```
      2 2 1 2
      46789
      56301
      67943
   +  11009
     182042
```

3. Subtract the following integers:

a. $123 - 45$

$$
\begin{array}{r}
\overset{\scriptscriptstyle 11\ 13}{12\!\!\!/3} \\
-\ 45 \\
\hline
78
\end{array}
$$

b. $45,678 - 34,908$

$$
\begin{array}{r}
\overset{\scriptscriptstyle 4\ 16}{45\!\!\!/678} \\
-\ 34908 \\
\hline
10,770
\end{array}
$$

c. $350,000 - 47,987$

$$
\begin{array}{r}
\overset{\scriptscriptstyle 4\ 9\ 9\ 9\ 10}{35\!\!\!/0\!\!\!/0\!\!\!/0\!\!\!/0} \\
-\ 47987 \\
\hline
302,013
\end{array}
$$

d. $358,000 - 47,987$

$$
\begin{array}{r}
\overset{\scriptscriptstyle 7\ 9\ 9\ 10}{358\!\!\!/0\!\!\!/0\!\!\!/0} \\
-\ 47987 \\
\hline
310,013
\end{array}
$$

e. $9,876 - 25$

$$
\begin{array}{r}
9876 \\
-\ 25 \\
\hline
9,851
\end{array}
$$

4. Multiply the following integers:

a. 56 × 78

$$
\begin{array}{r}
56 \\
\times\,78 \\
\hline
448 \\
\underline{3920} \\
4368
\end{array}
$$

b. 43,675 × 6

$$
\begin{array}{r}
43675 \\
\times\qquad 6 \\
\hline
262{,}050
\end{array}
$$

c. 469 × 372

$$
\begin{array}{r}
469 \\
\times\,372 \\
\hline
938 \\
32\,830 \\
\underline{140\,700} \\
174{,}468
\end{array}
$$

d. 144 × 233

$$
\begin{array}{r}
144 \\
\times\,233 \\
\hline
432 \\
4\,320 \\
\underline{28\,800} \\
33{,}552
\end{array}
$$

e. (21 × 34) × 55

```
        21              714
      × 34            ×  55
        84             3570
       630            35700
       714           39,270
```

5. Divide the following integers:

a. 1,625 ÷ 25

```
            65
      25)1625
         150
         125
         125
           0
```

b. 15,411 ÷ 33

```
           467
      33)15411
         132
         221
         198
         231
         231
           0
```

c. 74,539 ÷ 131

```
            569
      131)74539
          655
          903
          786
         1179
         1179
            0
```

d. 5,334 ÷ 9

$$\begin{array}{r} 592 \\ 9\overline{)5334} \\ \underline{45} \\ 83 \\ \underline{81} \\ 24 \\ \underline{18} \\ 6 \end{array}$$

e. 1900 ÷ 4

$$\begin{array}{r} 475 \\ 4\overline{)1900} \\ \underline{16} \\ 30 \\ \underline{28} \\ 20 \\ \underline{20} \\ 0 \end{array}$$

6. Evaluate the following:

a. 2^5

$$2 \times 2 \times 2 \times 2 \times 2 = 32$$

b. 1^9

$$1 \times 1 \times 1 \times 1 \times 1 \times 1 \times 1 \times 1 \times 1 = 1$$

c. 13^2

$$13 \times 13 = 169$$

d. 10^3

$$10 \times 10 \times 10 = 1000$$

e. $7^0 = 1$

7. Simplify the following expressions:

a. $(5 + 7) - 3 + 12 \div 2$

$$(5 + 7) - 3 + 12 \div 2 =$$

$$12 - 3 + 12 \div 2 =$$

$$12 - 3 + 6 =$$

$$9 + 6 =$$

$$15$$

ANSWERS

b. $(5 + 7 - 3 + 12) \div 12$

$$\frac{(5 + 7 - 3 + 12)}{12} =$$

$$\frac{(12 - 3 + 12)}{12} =$$

$$\frac{(9 + 12)}{12} =$$

$$\frac{21}{12} =$$

$$\frac{7}{4}$$

c. $8^2 + 2[41 - 3(6 - 4)]$

$$8^2 + 2[41 - 3(6 - 4)] =$$

$$8^2 + 2[41 - 3(2)] =$$

$$8^2 + 2[41 - 6] =$$

$$8^2 + 2[35] =$$

$$8^2 + 70 =$$

$$64 + 70 =$$

$$134$$

STUDY SIDEKICK

d. $56{,}274 - \{(12{,}000 + (3)(4{,}321) + 32675 \div 5)\}$

$56{,}274 - [12{,}000 + (3)(4{,}321) + 32{,}675 \div 5]$

$56{,}274 - [12{,}000 + 12{,}963 + 32{,}675 \div 5]$

$56{,}274 - [12{,}000 + 12{,}963 + 6{,}535]$

$56{,}274 - [31{,}498] =$

$24{,}776$

e. $56 \div 7 + (3^3 + 2)(14 - 9) \div 5(3^2 - 2^3)$

$$\frac{56}{7} + \frac{(3^3 + 2)(14 - 9)}{5(32 - 23)} =$$

$$\frac{56}{7} + \frac{(27 + 2)(14 - 9)}{5(9 - 8)} =$$

$$\frac{56}{7} + \frac{(29)(5)}{5(1)} =$$

$$\frac{56}{7} + \frac{145}{5} =$$

$8 + 29 =$

37

8. A new arena has 25,679 seats for which it can sell tickets at $15.00 and a total of 30,000 seats for which it can sell tickets.

   a. If it sells all the $15.00 tickets for a performance, how much will it gross for that performance?

   b. If it sells all the $15.00 tickets, and for $25 a ticket it sells half the remaining tickets for a performance, how much will it gross for that performance?

   c. If it sells 35% of the $25.00 tickets and 85% of the $15.00 tickets for a performance, how much will it gross for that performance?

Solution:

a. There are 25,679 tickets that are sold for $15.00 each. We simply multiply the number of tickets times cost of each ticket.

$$
\begin{array}{r}
25679 \\
\times\ \ \ 15 \\
\hline
128395 \\
\underline{256790} \\
385{,}185
\end{array}
$$

The arena will gross $385,185 when it sells all its $15.00 tickets for a performance.

b. This problem is a little more complicated. First we have to find out how many tickets can be sold at $25.00. Since we know the total number of seats is 30,000, and 25,679 tickets are purchased at $15, we can subtract to find the remainder of the seats available to be purchased for $25.

$$
\begin{array}{r}
30000 \\
- \ 25679 \\
\hline
4{,}321
\end{array}
$$

4,321 is how many tickets it is possible for the arena to sell at $25. They sold half of this amount of tickets. We can find out exactly how many by dividing.

$$
\begin{array}{r}
2160 \\
2\overline{)4321} \\
\underline{4} \\
03 \\
\underline{2} \\
12 \\
\underline{12} \\
01
\end{array}
$$

Notice that there is a remainder of 1. Since we can't sell half a ticket, we will round down to the nearest whole number, which is 2,160 tickets. These can be sold for $25.00 each, so this is just another multiplication problem:

$$
\begin{array}{r}
2160 \\
\times \ \ \ 25 \\
\hline
10800 \\
43200 \\
\hline
54{,}000
\end{array}
$$

Now we know the arena collects $54,000 for the $25 tickets. Remember we already found how much all the $15.00 tickets will gross in the first part of this problem. To find the total amount the arena made from the $25 and $15 tickets, we add $385,185 and $54,000.

$$
\begin{array}{r}
385,185 \\
+\ 54,000 \\
\hline
439,185
\end{array}
$$

The arena will gross $439,185 when it sells all the $15 tickets and half the remaining tickets for $25.

c. Okay, now, we know there are 25,679 tickets available at $15 and 4,321 tickets available at $25. For this problem, we have to find out how many tickets are sold at both prices. Then we'll multiply these amounts by their respective prices to find out how much the arena grossed. Let's start with the $15 tickets.

We know that 85% of the $15 tickets are sold. We also know that there are 25,679 tickets available at $15. First we find out what 85% of 25,679 is. Don't forget that to change a percent to a decimal you just move the decimal point 2 places to the left (and remember to round off since partial tickets don't get you jack).

$$
\begin{array}{r}
25679 \\
\times\ .85 \\
\hline
128395 \\
20\,54320 \\
\hline
21{,}827.15
\end{array}
$$

85% of the $15.00 tickets is 21,827 tickets. Now let's figure out the $25 tickets. So we'll find 35% of 4,321.

$$
\begin{array}{r}
4321 \\
\times\ .35 \\
\hline
21605 \\
129630 \\
\hline
1{,}512.35
\end{array}
$$

This means that 35% of the $25 tickets is 1,512 tickets. Now we just have to multiply the number of $15 tickets, 21,827, by $15, and the number $25 tickets, $251,512, by $25, like this:

$$
\begin{array}{r}
21827 \\
\times\ 15 \\
\hline
109135 \\
218270 \\
\hline
327{,}405
\end{array}
\qquad
\begin{array}{r}
4321 \\
\times\ 25 \\
\hline
21605 \\
86420 \\
\hline
108{,}025
\end{array}
$$

Next, we add these two totals together.

$$
\begin{array}{r}
32{,}7405 \\
+\ 10{,}8025 \\
\hline
435{,}430
\end{array}
$$

The final answer for this part is that the arena
will gross \$435,430.

9. A student is ordering pizzas and sodas for his study group.
The pizzas are \$6.00 a piece (six slices in a pizza) or \$1.50 per
slice. The sodas cost \$1.25 each. There are four men and five
women in the study group.

    a. If each man eats 4 slices of pizza, each woman eats 2
slices of pizza, and each student drinks 2 sodas, how
much will the bill come to (before tip)?

    b. If the tip is 15% of the bill, how much is the tip?

    c. If the students split the total bill (bill plus tip) equally,
how much should each person chip in?

Solution:

a. Okay, our first step is to find out how much pizza all the students will eat. We know that 4 men will each eat 4 slices of pizza and 5 women will each eat 2 slices of pizza. We'll multiply the number of slices each group will eat times the number of men and women eating.

$$4 \text{ men} \times 4 \text{ slices} = 16 \text{ slices}$$

$$5 \text{ women} \times 2 \text{ slices} = 10 \text{ slices}$$

Now we add up these two totals.

$$\begin{array}{r} 16 \\ + \underline{10} \\ 26 \text{ slices} \end{array}$$

They will eat 26 slices of pizza. Let's figure out how many whole pizzas and how many pizza slices they need. Because there are 6 slices in each pizza, 6 makes up the whole. The 26 slices of pieces are the part present. We can set this up as a fraction.

$$\frac{26}{6}$$

If we divide the denominator, the number of slices in each pizza, into the numerator, the amount of slices needed, we find the students need 4 whole pizzas at $6.00 each and 2 slices at $1.50 each. Let's figure out how much the pizza costs.

4 pizzas × $6 apiece = $24 in whole pizzas

2 slices × $1.50 each = $3 in single slices

Now just add up these two totals.

$$
\begin{array}{r}
24 \\
+\ 3 \\
\hline
27
\end{array}
$$

The pizza will cost $27.00. On to the sodas. Each person drinks 2 sodas. There's 4 men and 5 women, giving us 9 students in all. We just multiply the number of sodas each person drinks times the amount of students.

$$9 \times 2 = 18$$

The students will drink a total of 18 sodas. Each soda costs $1.25. We multiply these amounts together to find the total cost of all the sodas.

$$
\begin{array}{r}
1.25 \\
\times\ \underline{18} \\
1000 \\
\underline{1250} \\
22.50
\end{array}
$$

The soda comes to $22.50. The total bill (before the tip) will be the total cost of the pizza, $27.00, plus the total cost of the soda, $22.50.

$$
\begin{array}{r}
27.00 \\
+\ \underline{22.50} \\
49.50
\end{array}
$$

Shazzam. The total bill is $49.50.

b. All we have to do for this part is multiply 15%, or 0.15, time the total amount for the pizza and soda that we already determined.

$$
\begin{array}{r}
49.5 \\
\times\ \underline{.15} \\
2475 \\
\underline{4950} \\
7.425
\end{array}
$$

Rounding to the nearest penny, the tip will be $7.43.

c. To solve this part, we need to find the sum of the bill and the tip, then divide that sum by the total number of people (which we already found to be 9). First, we'll add the bill and tip.

$$
\begin{array}{r}
49.50 \\
+\ 7.43 \\
\hline
56.93
\end{array}
$$

Now that we know the total bill, we can divide this amount by the number of students.

$$
\begin{array}{r}
6.32 \\
9\overline{)56.93} \\
\underline{54} \\
29 \\
\underline{27} \\
23 \\
\underline{18} \\
5
\end{array}
$$

Each student has to pay a mere \$6.32 and $\frac{5}{9}$ cents each. We can round this up to \$6.33 each.

10. A student is checking her car for gas mileage performance during a recent three-day trip. She noted her beginning and ending odometer readings as well as the gas required each time she stopped to fill up the gas tank.

STUDY SIDEKICK

Starting Odometer	Ending Odometer	Gallons of Gas Used
12,500.9	12,832.4	13.9
12,832.4	13,057.6	9.5
13,057.6	13,345.6	12.3

a. Gas mileage is found by dividing the number of miles driven by the number of gallons of gas used during the drive. Find the student's gas mileage for the three different occasions when she stopped for gas shown in the table above.

b. If she purchased gas at an average of $1.46 per gallon, what was her cost for the gas used during the trip?

Solution:

a. We want to find the gas mileage for each time she stopped. But first, we must figure out the number of miles driven by subtracting the odometer readings (smaller from larger). Then, we divide the number of of miles driven by the number of gallons of gas used during the drive. That'll give us her gas mileage. We'll start by looking at the first time she stopped, which is the information on the first line of the table. We subtract the odometer readings.

$$
\begin{array}{r}
{}^{1\,14} \\
1283\cancel{2}.\cancel{4} \\
-\ \underline{12500.9} \\
331.5
\end{array}
$$

Now we divide the gallons of gas she's used by the 331.5 miles she's driven.

```
              23.84
      139)3315.00
          278
          535
          417
         1180
         1112
          680
          556
          124
```

Her gas mileage the first time she stopped was about 23.8 miles per gallon.

We do the other two stops the same way, subtracting the odometer readings.

```
        2 10
      13057.6
    − 12832.4
        225.2
```

Now find the gas mileage.

```
        23.70
95)2252.00
   190
   352
   285
   670
   665
    50
     0
    50
```

Her mileage is about 23.7 miles per gallon in this second part. One more line left in the table. Again, we subtract to find the miles driven.

```
   2 13 15
 13345.6
-13057.6
   288.0
```

Our final operation is to find the mileage for that part of the trip.

```
         23.41
123)2880.00
    246
    420
    369
    510
    492
    180
    123
     57
```

Her mileage was about 23.4 miles per gallon for this last part of the trip. We've finished the first part of the problem.

b. The second part asks us to find the total cost of the gas used during the trip. We're given the average price of a gallon of gas. Since the gas price is an average, all we have to do is multiply the total number of gallons she used by this average price. First, we'll add up the gallons of gas she used.

$$
\begin{array}{r}
{\scriptstyle 1\,1} \\
13.9 \\
9.5 \\
+\ \underline{12.3} \\
35.7
\end{array}
$$

Now that we know the total gallons used, we just multiply this amount by $1.46, the average price of each gallon.

$$
\begin{array}{r}
35.7 \\
\times\ \underline{1.46} \\
2142 \\
14280 \\
\underline{35700} \\
52.122
\end{array}
$$

This means that she used 35.7 gallons and it cost her $52.12 (rounding to the nearest cent) to pay for all the gas.

11. Write the following numbers in scientific notation:

a. 647

$$6.47 \times 10^2$$

b. 2,345,543,247,000

$$2.345543247 \times 10^{12}$$

c. 0.00000563

$$5.63 \times 10^{-6}$$

d. 8,011,432.007

$$8.011432007 \times 10^6$$

e. 100

$$1 \times 10^2$$

12. Reduce the following fractions to their lowest terms:

a. $\dfrac{25}{75}$

$$\frac{25}{75} = \frac{25 \div 25}{75 \div 25} = \frac{1}{3}$$

ANSWERS

b. $\dfrac{130}{26}$

$$\frac{130}{26} = \frac{130 \div 26}{26 \div 26} = \frac{5}{1} = 5$$

c. $\dfrac{17}{51}$

$$\frac{17}{51} = \frac{17 \div 17}{51 \div 17} = \frac{1}{3}$$

d. $\dfrac{210}{9240}$

$$\frac{210}{9240} = \frac{210 \div 210}{9240 \div 210} = \frac{1}{44}$$

e. $\dfrac{105}{1155}$

$$\frac{105}{1155} = \frac{105 \div 105}{1155 \div 105} = \frac{1}{11}$$

13. Solve the following fractions, making sure that all answers have been reduced to their lowest terms:

a. $\dfrac{1}{4} + \dfrac{5}{7}$

4 and 7 have an LCD of 28:

$$\frac{1 \times 7}{4 \times 7} + \frac{5 \times 4}{7 \times 4} =$$

$$\frac{7}{28} + \frac{20}{28} =$$

$$\frac{27}{28}$$

b. $\dfrac{1}{3} - \dfrac{12}{51} + \dfrac{15}{18}$

3, 51 and 18 have an LCD of 306:

$$\frac{1 \times 102}{3 \times 102} - \frac{12 \times 6}{51 \times 6} + \frac{15 \times 17}{18 \times 17} =$$

$$\frac{102}{306} - \frac{72}{306} + \frac{255}{306} =$$

$$\frac{102 - 72 + 255}{306} =$$

$$\frac{285}{306} =$$

$$\frac{95}{102}$$

c. $4 + \dfrac{2}{5} + \dfrac{1}{12}$

1, 5, and 12 have an LCD of 60:

$$\frac{4 \times 60}{1 \times 60} + \frac{2 \times 12}{5 \times 12} + \frac{1 \times 5}{12 \times 5} =$$

$$\frac{240}{60} + \frac{24}{60} + \frac{5}{60} =$$

$$\frac{240 + 24 + 5}{60} =$$

$$\frac{269}{60} =$$

$$4\frac{29}{60}$$

d. $\dfrac{3}{4} - \dfrac{21}{32} + \dfrac{1}{17}$

4, 32, and 17 have an LCD of 544:

$$\frac{3 \times 136}{4 \times 136} - \frac{21 \times 17}{32 \times 17} + \frac{1 \times 32}{17 \times 32} =$$

$$\frac{408}{544} - \frac{357}{544} + \frac{32}{544} =$$

$$\frac{408 \quad 357 + 32}{544} =$$

$$\frac{83}{544}$$

14. Multiply the following:

a. $\dfrac{2}{5} \times \dfrac{3}{7}$

$$\dfrac{2 \times 3}{5 \times 7} =$$

$$\dfrac{6}{35}$$

b. $\dfrac{3}{7} \times \dfrac{6}{4} \times \dfrac{1}{11}$

$$\dfrac{3 \times 6 \times 1}{7 \times 7 \times 11} =$$

$$\dfrac{18}{308} =$$

$$\dfrac{9}{154}$$

c. $7\dfrac{2}{3} \times \dfrac{5}{8}$

$$\dfrac{23}{3} \times \dfrac{5}{8} =$$

$$\dfrac{115}{24} =$$

$$4\dfrac{19}{24}$$

d. $\dfrac{1}{4} \times \dfrac{6}{73} \times \dfrac{73}{18}$

$$\dfrac{1 \times 6}{4 \times 18} =$$

$$\dfrac{6}{72} =$$

$$\dfrac{1}{12}$$

Notice that we cancelled out the 73 in the top and bottom of the fraction before we multiplied. It doesn't change our answer and it simplifies the multiplication process.

15. Divide the following:

a. $\dfrac{2}{3} \div \dfrac{1}{5}$

$$\dfrac{2}{3} \times \dfrac{5}{1} =$$

$$\dfrac{10}{3} =$$

$$3\dfrac{1}{3}$$

b. $\dfrac{5}{8} \div \dfrac{8}{31}$

$$\dfrac{5}{8} \times \dfrac{31}{8} =$$

$$\dfrac{155}{64} =$$

$$2\dfrac{27}{64}$$

c. $\dfrac{11}{4} \div \dfrac{3}{4}$

$$\dfrac{11}{4} \times \dfrac{4}{3} =$$

$$\dfrac{44}{12} =$$

$$3\dfrac{2}{3}$$

d. $\dfrac{16}{25} \div \dfrac{8}{9}$

$$\dfrac{16}{25} \times \dfrac{9}{8}$$

$$\dfrac{144}{200} =$$

$$\dfrac{18}{25}$$

e. $(2 + \frac{3}{8}) \div (1 + \frac{4}{7})$

$$(\frac{16}{8} + \frac{3}{8}) \div (\frac{7}{7} + \frac{4}{7}) =$$

$$\frac{19}{8} \div \frac{11}{7} =$$

$$\frac{19}{8} \times \frac{7}{11} =$$

$$\frac{133}{88} =$$

$$1\frac{45}{88}$$

When you get down to it, any time the ant generals can have an end-of-the-day meeting and say they've only lost 60 ants or so that day... It's a good day!

– Igor Torgeson

16. Find the decimal and percent forms of the following numbers:

a. $\frac{5}{8}$

$$\frac{5}{8} = 0.625 = 62.5\%$$

b. $\frac{2}{9}$

$$\frac{2}{9} = 0.22222 = 22.222$$

c. $1\frac{2}{7}$

$$1\frac{2}{7} = 1.2587 = 125.87\%$$

d. 42

$$42 = 42.00 = 4200\%$$

e. $\dfrac{11}{6}$

$$\dfrac{11}{6} = 1.8333 = 183.33\%$$

17. If a student has 40 pages of notes from his math class and 15 pages of notes from his history class, what is the ratio of math notes to history notes?

Solution:

The ratio of math notes to history notes is 40:15, which reduces to 8:3. We see that there are 2.67 times as many math notes as there are history notes (because $\dfrac{8}{3}$ is 2.67).

18. What is 62.5% of 1,142, if you round the answer to the nearest units place?

Solution:

$$
\begin{array}{r}
1142 \\
\times\ 0.625 \\
\hline
5710 \\
22840 \\
\underline{685200} \\
713.750
\end{array}
$$

When we round 713.75 to the nearest units place, the answer is 714.

19. Deviant University has a freshman class of 4,580 students. If 30% of the students want to take freshman English at 8 a.m. and there are five classes taught at that hour with no more than 100 students in each, how many students who want to get into that early English class will not be able to?

Solution:

Okay, our first step is to find out how many people want to get into this class. We know 30% of the freshman class want to get in. We'll multiply this percent times the number of students, 4,580.

$$
\begin{array}{r}
4580 \\
\times\ \underline{.3} \\
1374.0
\end{array}
$$

We know 1374 students want to take the class at this time. Now, we'll find out how many can take a class at this time. There can be only 100 students in each class and there are 5 classes taught at this time. As you guessed, this is a multiplication problem.

$$
\begin{array}{r}
100 \\
\times\ \underline{5} \\
500
\end{array}
$$

We can subtract the number of students who can take a class at this hour from the number of students who want to take a class at this time by simply subtracting.

$$
\begin{array}{r}
1374 \\
-\ 500 \\
\hline
874
\end{array}
$$

Soooo... 874 freshmen will not be able to get into the early English class.

20. Ms. Dina Soar, a paleontologist, has uncovered a total of 840 fossilized bones in her latest dig. If she has identified that 10% are from velociraptors and $\frac{1}{4}$ are from pteranodons, how many does she have left to identify? What is the ratio of the identified bones to the unidentified bones?

Solution:

We can do this problem in two different ways. The first way is to find out how many bones are from velociraptors (by finding 10% of 840), then find out how many bones are from pteranodons (by finding 25% of 840), then adding those numbers together and subtracting THEIR sum from 840.

The second way is to notice that the problem asks for how many unidentified bones there are. We could just say that, if Dina knows what 10% are and she knows what 25% are, then

she can identify 10% plus 25%, or 35% of what the bones are. All we have to figure out is how many bones 100% - 35%, or 65%, are. To do this, we must find 65% of 840.

$$
\begin{array}{r}
840 \\
\times\,.65 \\
\hline
4200 \\
\underline{50400} \\
546.00
\end{array}
$$

Dina has not identified 546 bones. This means the ratio of identified to unidentified bones is (840 − 546) : 546, which is 294 : 546. We can reduce this to 7 : 13. This means she has a little over half the bones left to identify.

21. Dr. James Pearson, famous astronomer, has discovered a new heavenly body that is approximately 0.25 light-year away. If a light-second is equal to 186,000 miles and there are 60 light-seconds in a light-minute and 60 light-minutes in a light-hour and 24 light-hours in a light-day and 365.6 light-days in a light-year, how many miles away is Dr. Pharr's heavenly body (use scientific notation and round your answer to three decimal places)?

STUDY SIDEKICK

Solution:

This is really just a conversion problem, where we are asked to convert from one unit to another (just like we did in the problem about pizzas and sodas). All we have to do is multiply all the information we're given above, since we're just converting measurements.

$$0.25 \times 365.6 \times 24 \times 60 \times 60 \times 186,000 \text{ miles}$$

First, we'll multiply the number of light-years away the heavenly body is times the number of light-days in a light-year.

$$
\begin{array}{r}
365.6 \\
\times\ .25 \\
\hline
18280 \\
\underline{73120} \\
91,400
\end{array}
$$

We know that the heavenly body is 91,400 light-days away. We then go through each of these multiplication operations to reach our final answer. Here's what our answer is after we multiply all these figures out.

$$1,468,834,560,000$$

Let's put this huge number into scientific notation.

$$1.469 \times 10^{12}$$

ANSWERS

22. Two years ago, there was 1 coffee bar for every 4 blocks in Washington, D.C. Now, there are 3 coffee bars on every block of that same city. What is the ratio of the coffee bars today to the coffee bars 2 years ago?

Solution:

2 years ago, the ratio of coffee bars to blocks was 1 : 4 or $\frac{1}{4}$. Now, the ratio of coffee bars to blocks is 3 : 1 or $\frac{3}{1}$. We can say that the ratio of coffee bars now to coffee bars 2 years ago is the same as 3 : 1 / 1 : 4.

$$-\frac{3}{1} \div \frac{1}{4}$$

$$-\frac{3}{1} \times \frac{4}{1}$$

$$\frac{12}{1}$$

The ratio is 12 : 1, which tells us that there are 12 times more coffee bars now than there were 2 years ago.

23. A student working his way through college was paid at an hourly rate of $7.62. After he had gained some experience, he received a merit raise of 5%. Two months later, he, like all the other employees, received a cost of living raise of 3.1%. What was his hourly rate after both raises?

STUDY SIDEKICK

Solution:

This problem is very straightforward. We find out how much he was making after his merit raise by multiplying his 5% pay increase by what he was making before the raise ($7.62).

$$
\begin{array}{r}
7.62 \\
\times\ .05 \\
\hline
.3810
\end{array}
$$

He received a $0.38 pay increase with the first raise. We'll add this onto his previous hourly wage.

$$
\begin{array}{r}
7.62 \\
+\ .38 \\
\hline
8.00
\end{array}
$$

Now we have to figure in his 3.1% cost-of-living raise.

$$
\begin{array}{r}
8.00 \\
\times\ .031 \\
\hline
800 \\
24000 \\
\hline
.24800
\end{array}
$$

He received a $0.25 increase with his second wage increase. We'll add this onto his previous hourly wage.

$$
\begin{array}{r}
8.00 \\
+\ .25 \\
\hline
8.25
\end{array}
$$

His hourly rate of pay after both raises was $8.25.

GLOSSARY

addition - An operation performed on two numbers at a time that results in combining both numbers into a third number, such as $1 + 2 = 3$.

arithmetic - A field of mathematics that studies real numbers and the operations of addition, subtraction, multiplication, and division, among others. It is basic to the study of mathematics.

borrow - A technique used in subtraction to permit the subtraction of a numeral that is larger than the one it is subtracted from. It gets its name from the ancient calculators called counting boards, in which beads were borrowed from a higher place holder and put in a lower one to let the operation of subtraction be done.

calculator - Technology's gift to math students everywhere. Calculators were used as far back as the Roman Empire, when they were made of small beads that could be moved in grooves to perform addition, subtraction, and multiplication. Division was done mostly as a series of subtractions. The ancient Chinese strung their small beads on rods to make a calculator known to us as an abacus; the ancient Japanese used a similar instrument. Ancient Russians, Turks, and Armenians all developed characteristic "counting boards" to help them perform quick, accurate computations.

carry - A method used in addition in which you carry an amount to the adjacent digit place one place higher in value. When addition was done with an abacus or other counting board, the stone or counter was actually lifted from one place and put in the next highest place.

coffee break - A short period for rest and refreshments.

composite number - A number that has more factors than itself and 1. For example, the number 20 is a composite number because it has 1, 2, 4, 5, 10, and 20 as factors. Any number that is not a prime number is a composite number.

decimal - A number that is written in the scale of tens.

decimal point - The symbol "." that is used in the U.S. to mark the place in a numeral that is between the whole number and the fraction of the rest of the number. For example, in 10.25, the decimal point marks the place between the whole number 10 and the decimal 0.25 (also known, in fraction form, as $\frac{1}{4}$).

denominator - In a fraction, the denominator is the number below the bar. 4 is the denominator in the fraction $\frac{3}{4}$.

digit - Any of these numerals: 0, 1, 2, 3, 4, 5, 6, 7, 8, 9.

dividend - A number that is divided by another number. For example, if you divide 6 into 144, the dividend is 144.

division - A mathematical operation performed on two

numbers. The inverse of multiplication. In ancient times, division was rarely used except when the divisor was very small. Different methods were used by different cultures. Long division, still in use today, showed up in print a year before Columbus discovered America.

divisor - A number that is divided into another number. For example, if you divide 6 into 144, the divisor is 6.

exponent - A number that is written above and to the right of another number (the base) to indicate repeated multiplication of the base number times itself. An exponent is also called a "power." In the expression 5^2, the exponent is 2 and the base is 5.

factor - A factor is a number that can be divided evenly (that is, with no remainder) into some other number. A number can have many factors or just two factors. For example, the number 12 has six factors, 1, 2, 3, 4, 6, and 12, while the number 3 has only two factors, 1 and 3.

fraction - A fraction is a number that is in the form of an indicated division of two integers, such as $\frac{1}{3}$ or $\frac{3}{2}$ or $\frac{1}{1000000001}$ or $\frac{-3}{5}$ or $\frac{-15}{1}$.

handball - A game played by two or four players against a single wall or in a walled court in which the players use their hands to hit the ball.

STUDY SIDEKICK

hundreds place - In a numeral, this is the third digit place from the right. For example, in the numeral 1,200, the 2 is occupying the hundreds place (and stands for 200). Likewise, in the numeral 1200.056, the 2 is STILL occupying the hundreds place.

hundredths place - In a numeral, this is the second digit place to the right of the decimal point. For example, in the numeral 1200.056, the 5 is occupying the hundredths place.

improper fraction - A fraction having a larger numerator than denominator, such as $\frac{22}{7}$, is said to be an improper fraction.

infinity - A mathematical concept that has no real equivalent on Earth. Something that has no end, that cannot be described by a real number, and that is either very, very, very, very, very, very large or very, very, very, very, very, very small.

integer - The set of all positive and negative whole numbers, including zero.

interest - Percent charged on how much you borrowed. Multiply percent times amount borrowed to find how much interest is owed.

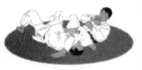

jujitsu master- A person of consummate skill in the art of weaponless fighting, in which one tries to disable an opponent with holds, throws, and paralyzing blows.

lowest common denominator (LCD) - The smallest number divisible by all of the denominators in the expression to be evaluated. It will never be smaller than the largest denominator in the expression. For example, the lowest common denominator for $\frac{1}{2} + \frac{1}{3}$ would be 6, since both 2 and 3 divide into 6 evenly.

Matagorda Bay - An inlet of the Gulf of Mexico 30 miles long in southeast Texas. Home of some great speckled trout fishing.

minus - The name for the symbol "$-$" that means "subtract the number after me from the number before me." When you see this word in a word problem, it means subtract. For example, "four minus three" means $4 - 3$.

mixed number - A number that consists of a whole number and a fraction, like $3\frac{1}{4}$. It's the number you get when you add the whole number and the fraction together.

multiplication - A mathematical operation performed on two numbers at a time, which allows a faster calculation of the operation of addition. Any multiplication operation can be performed as a series of additions. For example, $3 \times 3 = 9$ gives us the same answer as $3 + 3 + 3 = 9$. Multiplication is a shortcut for addition (especially for large numbers) and was probably the third mathematical operation to be invented by ancient cultures.

number - A number is a quantity. It is a very basic concept that has had very brilliant and highly educated people sweating over a proper definition for centuries.

number line - A visual aid in mathematics used by beginners and experts alike. It is a straight line, either horizontal or vertical, and evenly spaced divisions marked on it, with the positive numbers extending to the right (or upward, if the line is vertical) and the negative numbers extending to the left (or downward).

numeral - People often confuse this with a number. A numeral is just a particular symbol, like 4, that stands for the number four. It means a specific graphic representation, but not the number itself.

numerator - The numerator of a fraction is the number above the bar. For example, in the fraction $\frac{2}{4}$, the numerator is 2.

operation - A defined mathematical action, typically performed on one or two numbers, such as addition, subtraction, multiplication, and division.

order of operations - A set of priorities for doing mathematical operations. This order never changes. Parentheses, brackets, or some other grouping symbol, are performed first. Exponents are the next operation, then any multiplication or division operations, and finally, all additions and subtractions are performed. When you work out a bunch of indicated operations, you should ALWAYS start on the left and proceed to the right. (PEMDAS!)

percent - A part of a whole, expressed in hundredths. For example, 20% means $\frac{20}{100}$ and 100% means $\frac{100}{100}$ or 1.

plus - The name of the symbol "+" that means "add together the number before me and the number after me." When you see this word in a word problem, it means add. For example, "one plus three" means 1 + 3.

popsicle™ - Flavored water frozen into the shape of a rectangle, a rocket ship (or anything else for that matter) then stuck on a stick.

prime number - A positive number that can only be divided evenly (that is, with no remainder) by itself and 1. For example, 17 is a prime number because no numbers divide into it evenly except for 17 and 1. Every number that is not a prime number is a composite number.

product - A product is the result of a multiplication operation. Another way to say that 2 × 3 *equals* 6 is that its product is 6.

proper fraction - A proper fraction is one whose numerator is a smaller number than its denominator. $\frac{1}{4}$ is a proper fraction, as is $\frac{2}{5}$, $\frac{1300}{1301}$, etc.

quotient - A quotient is the result of a division operation. If you divide 12 by 6, the quotient is 2. If you multiply the quotient by the divisor and then add in the remainder, you should get the dividend. If not, stop everything and rework the problem very carefully, since there is a mistake somewhere.

ratio - A relationship between the quantity of one thing and the quantity of one or more other things (in similar units of measure). For example, if a recipe calls for one teaspoon of salt and one tablespoon of flour, you can say that the ratio of salt to flour is one teaspoon to one tablespoon. Or you can convert the measures, so that one tablespoon of flour is equal to three teaspoons of flour and then you can say the ratio of salt to flour is 1:3 and leave out the units of measurement.

rational - In mathematics, a rational number is any number that can be put in the form of a fraction.

remainder - A remainder is what is left after you divide one number by another number. If you divide 13 by 4, the answer is 3, remainder 1. Remainders are usually expressed as fractions or decimals, like this: $13 \div 4 = 3\frac{1}{4} = 3.25$.

rounding - This is a standardized technique for making numbers a bit smaller or larger (and less exact), so they are easier to work with. It is sometimes called rounding off, because you are shortening the length of numbers when you round. For example, to round 1024.66666666666 to the nearest tenth, you look at the hundredths place, and since it's more than 5, you round to 1024.7. See? It's much shorter, much easier to work with.

rounding down - This is a variety of rounding that specifically means that the number you end up with is a tiny bit smaller than the number you started with. For example, if we round the number 345.2 to the nearest unit and say it is about 345, we have rounded down.

GLOSSARY

rounding up - This is the opposite of rounding down. It means that the number you end up with is a bit bigger than the number you started with. For example, if we round the number 999.6 to 1,000, we have just rounded up. If you are herding together a group of cattle, you are also rounding up, but this operation is not usually done in math classes (except in Wyoming).

scientific notation - This is a convention that lets us write extremely large and extremely small numbers without using all the paper from a whole tree. The scientific notation for 12,000,000,000,000,000,000,000,000,000,000 is 1.2×10^{28}.

somersault - A roll forwards or backwards in which a person brings her feet over her head in a complete revolution until she's on her feet again.

square - In mathematics, to square something means to multiply it by itself once. It is indicated by the exponent 2, like this: $3^2 = 3 \times 3 = 9$.

square dance - A dance involving four couples who form a square.

subtraction - A mathematical operation performed on two numbers where one number is taken away from another number. Subtraction is the inverse of addition. Subtraction was probably the second mathematical operation invented by ancient cultures.

> First off, what kind of guy keeps a toothbrush in his pocket? And what kind of guy keeps droppin' it on the floor? I mean, keep it in your desk for goodness sake.
>
> – C. Alan Canant

tens place - In a numeral, this is the second place to the left of the decimal point. For example, in the numeral 1234, the 3 is occupying the tens place (and stands for 30). Likewise, in the numeral 1234.056, the 3 is STILL occupying the tens place.

tenths place - In a numeral, this is the first place to the right of the decimal point. For example, in the numeral 1234.056, the 0 is occupying the tenths place.

thousands place - In a numeral, this is the fourth place to the left of the decimal point. For example, in the numeral 12,567, the 2 is occupying the thousands place (and stands for 2,000). Likewise, in the numeral 12,567.056, the 2 is STILL occupying the hundreds place.

thousandths place - In a numeral, this is the third place to the right of the decimal point. For example, in the numeral 1200.056, the 6 is occupying the thousandths place.

tripe - The stomach tissue of an animal (usually of an ox) often used in stews.

units place - In a numeral, this is the first place to the left of the decimal point, if the numeral represents an integer. For example, in the numeral 987, the 7 is occupying the units place. Likewise, in the numeral 987.056, the 7 is STILL occupying the units place.

GLOSSARY

uniths place - There is no such thing. Units ONLY exist as integers, or whole numbers.

zany - An absurdly ludicrous act. Also, one who acts the buffoon to amuse others.

STUDY SIDEKICK

PERSONAL NOTES